CW00678551

Firefighters
& the Blitz

Firefighters
& the Blitz

Francis Beckett

MERLIN PRESS

© Francis Beckett, 2010

First published in 2010 by
The Merlin Press Ltd.
6 Crane Street Chambers
Crane Street
Pontypool
NP4 6ND
Wales

www.merlinpress.co.uk

ISBN. 978-0-85036-673-0

British Library Cataloguing in Publication Data
is available from the British Library

We have endeavoured to obtain permission from copyright holders of all
works reproduced in the book. We apologise if we have failed.

All rights reserved. No part of this publication may be reproduced, stored in a
retrieval system, or transmitted, in any form or by any means, electronic,
mechanical, photocopying, recording or otherwise, without the prior
permission of the publisher.

Printed in Malta on behalf of LPPS Ltd. Wellingborough, Northants, UK

Contents

Acknowledgements

The initiative for producing an account of the work of firefighters during the Blitz came from the Executive Council of the Fire Brigades Union. They put the union's resources behind it. I want to thank the union's general secretary Matt Wrack, the president Mick Shaw and other officials for their support. A number of members of the union's staff also assisted in numerous ways. Without them all, this book would not have been possible. They have been a pleasure to work with.

Matt Wrack also commissioned two talented film-makers, Alan Miles and Winstan Whitter, to make a film on the subject, and I have worked closely with both of them, often going out and doing interviews with them, and sometimes just using the product of an interview they did, even though I wasn't present. They too were a pleasure to work with, and generous with the products of their research.

With a tight timescale, I needed help to get the book written in time for the seventieth anniversary of the start of the Blitz, and was fortunate that Tony Russell and Peta Van Den Bergh, both considerable authors in their own right, were able to find some time to help me.

Jane Rugg at the London Fire Brigade Museum and Chris Coates, who runs the TUC archive at London Metropolitan University, gave me the run of their collections and helpfully guided me through them. Terry Segars gave me access to his interesting interviews with wartime firefighters, including his interview with John Horner. Stephanie Maltman, founder and co-chair of Firemen Remembered, gave generously of her vast store of knowledge and saved me from several mistakes.

I needed to go in search of firefighters who had fought the Blitz. For help with this, my thanks to the National Association of Retired Firefighters, who advertised my quest in their journal; and also to the University of the Third Age, whose magazine *U3A News* (which I happen to edit) is the best place I know to find people who have seen some of the crucial events of the last seventy years or so.

It did not take me long to realise that, since many of the firefighters during the Blitz were over military age at the time, few are left alive now. Alan, Winstan and I did get to one or two, and we also found that their sons and daughters, who were children during the Blitz, had many memories worth preserving.

I'm grateful for the stories and insights that came from our interviews with, or letters or emails from, Frank Bailey, Marion Bloom, Eric Bulmer, Merle Cartwright, Doreen & Don Chinnery, John Cunnington, John Dyson, Rosemary Graves, Frank Harbud, Frederick (Jack) Horner, Margaret Hughes, Angus Idle, Dr Eric Midwinter, Lyn Nixon (who also loaned me her precious copy of the now very rare 1945 Stephen Spender book on the Blitz), Roy Sargeant, Stacey Simkins (who also let me see his entertaining memoirs), Clive Skippins and Professor Linden West.

Thanks also to Anthony Zurbrugg and Adrian Howe of Merlin Press, who have published this book in association with the FBU.

Francis Beckett
June 2010

Introduction

In 2002, during our long and bitter struggle over pay, I arranged for a BBC journalist to spend some time with firefighters at an east London fire station. He told me he had never experienced such an intense sense of solidarity among a group of working people.

The fire service is like that. Working in relatively small teams for long periods, sharing jokes, sharing work and sharing risks gives a unique character to firefighting. This solidarity and camaraderie is reflected in this book and it is still there today. It still gives a special character to fire stations and other fire service workplaces all over the UK – indeed all over the world.

It also gives a special character to the Fire Brigades Union. We stick together and look after each other on the fireground – and we also stick together and look after each other as workers and trades unionists. And some people hate that. Regrettably, today there are those in positions of authority who resent the 'watch culture', seeing it as subversive and threatening. Thankfully they have not been successful in eliminating it and I hope they never will.

The Second World War was of huge significance to the fire service in the UK and to the Fire Brigades Union. The challenges posed by mass bombing prompted debates about the best way to arrange an effective and efficient firefighting organisation. The pre-war fire service was hugely fragmented, famously using different and incompatible equipment in different parts of the country. The challenges of wartime forced a radical re-think about what was necessary. The service which emerged after the end of the war was very different. It was a creation of the post-war desire to do things better, to plan better – and to ensure that the role of working people would be recognised. The Fire Brigades Union helped to make sure of that.

The war presented huge challenges to the union. The FBU could easily have been destroyed. The bombing threat meant that the government rapidly recruited thousands of new people to prepare to fight the coming fires. The influx of these Auxiliary Fire Service (AFS) firefighters would have swamped the existing professionals had the union not taken the bold and progressive stance of seeking to overcome divisions by recruiting AFS men and women into the union.

This decision, taken after much heated debate, was a turning point for the union and for the firefighting profession. It allowed the FBU to grow rapidly, ensuring its survival and increased strength and influence. This increased strength and the key role the FBU played in efforts to fight the Blitz meant that the FBU could play a central role in the post-war debates and the creation of the new post-war fire service.

That post-war service was a remarkable creation. The structures lasted more than fifty years and ensured that the UK fire service became respected throughout the world.

The years after 2003 have seen the unfolding of a process called 'modernisation' of the service; an over-used term widely discredited among firefighters. Under this process, anything existing before 2004 has been seen as belonging to the 'bad old days' while anything which has happened since (no matter how daft) is seen as modern and progressive. The truth is that the achievements of the wartime and post-war fire service still provide the basis for much of what happens in fire services today – despite the huge effort to pretend otherwise.

The creation of the AFS began a process which would eventually completely transform firefighting in the UK. Nationalisation of the service in 1941 took this to the next stage. Nationalisation was undertaken to ensure that resources – equipment and people – could rapidly be provided where most needed. Today, as a new government embarks on a period of 'localism', the FBU has asked how such an approach will adequately deal with challenges which go beyond the local – such as widespread flooding or terrorist attack. The scale of the challenge may be different to that faced in wartime but the tension between the local and the national remains.

Nationalisation also ensured standardisation of procedures and practices through measures such as the production of a national *Manual of Firemanship* and *Drill Book*. Firefighters rely on their skills, knowledge and training as well as on appliances, equipment and procedures. These are essential to firefighters' safety and to providing an effective and efficient service to the community. They are essential to ensure that we can put out a fire with as little risk and as little damage as possible. Over the past five years the FBU has argued that an effective and efficient fire and rescue service – even one delivered locally – needs to ensure that the best professional standards (appliances, equipment and procedures) are applied across the UK. We continue to fight against the fragmentation of our profession and our service arising from the latest political fad. Different times – same battles!

This book tells of historic events. But more importantly it tells the story of the people involved – the firefighters who fought those huge fires during the mass bombing of Britain. The stories of those people will strike a chord with people today – and perhaps especially with firefighters. You may share John Horner's anger at the treatment of Harry Errington and you may hold back tears reading of the shared grave of Benjamin Chinnery and Herbert Blundell – solidarity and teamwork in death.

Many of those involved in the events described here and who survived the war are now dead and unable to contribute directly. In many cases, their stories are kept alive by their families who have remembered the tales, stored the press articles and kept the memorabilia. They have done a service to the memory of their loved ones but also to all the firefighters involved.

We all hope that we never see bombings and fires on the

scale recounted in this book but during my five years as general secretary of the FBU we have seen some very major and diverse challenges. The London bombings of 2005 ... the huge Buncefield fire ... the floods of 2007 ... all of these have once again pushed firefighters onto the front line of protecting public safety. All of these have again demonstrated the immense flexibility and resourcefulness that is a feature of the firefighting profession. Firefighters take on these new challenges and are proud to do so. We merely believe that in return we have a right to the best resources – equipment, procedures, training and people – to be able to carry out the job professionally. And, just as John Horner and his generation recognized, we too have the right to decent pay, pensions and conditions of service.

Reading this book also reminded me of another feature of our recent history. We have lost far too many colleagues and attended far too many fire service funerals over the past five years. The grief felt by the families and colleagues of those killed in the recent past is on a different scale but no less devastating than that felt by friends and families during the war years. I hope that this book will remind readers of the risks faced by firefighters – then and now – so that our communities can be made safer. I hope it will remind us all that our fire and rescue service is indeed a service to be proud of – and a service worth defending.

Matt Wrack
General Secretary of the Fire Brigades Union
June 2010

Chapter 1

To 3 September 1939
Preparations for war

The year Hitler came to power, 1933, was also the year the British Home Office began to think about how they would cope with bombing, which was clearly going to be a feature of the next war. But it was not until 1937, when German bombers supporting the fascists in the Spanish Civil War destroyed the defenceless town of Guernica, that people started to realise just how dreadful it could be.

Two thousand civilians died in Guernica, most of them in the fire and smoke that followed the bombings. Nothing like it had ever been seen before, and Guernica convinced many people that a civilian population could be bombed into submission. What might such an attack do to London's four million inhabitants, its packed and inflammable warehouses, its maze of narrow streets by the river, its teeming slums where thousands lived cramped together? The government would surely be forced to surrender, or the people would take to the streets and overthrow it.

But Spain, if anyone had been listening, also showed how a city might defend itself. Ramon Perera, a Catalan engineer, supervised the building of 1,400 public shelters in Barcelona. They proved a great success. No one was killed in the shelters despite frequent heavy air raids on the city. The British structural engineer Cyril Helsby went to Barcelona in December 1938 on a fact-finding visit sponsored by the Labour Party, and persuaded the British secret services to help Perera reach Britain so that he might help prepare Britain's defences.

But the National Government (effectively a Conservative government) that was in power from the summer of 1931 when the Labour government fell until the summer of 1940, did not see war as inevitable. In the second half of the 1930s, under Stanley Baldwin and then Neville Chamberlain, it was slow to re-arm, and slow to prepare the nation's cities to defend themselves.

Perera's public shelters were not built. Eventually families were given Anderson shelters (named after the Home Secretary, Sir John Anderson) instead. This, as Stephen Spender wrote immediately after the war, "overlooked the fact that in the majority of homes there was no room for an Anderson shelter."

If adequate protection for the people could not be built, could the fires be effectively fought? No: Britain's fire services were not in a fit state to take on a nationwide emergency such as the one that had destroyed Guernica and many of its inhabitants. There was no national fire service. Instead there were 1,600 independent fire brigades, controlled by local councils. The biggest, the London Fire Brigade, had full-time crews with 106 pumping appliances. Some of the smallest were controlled by parish councils and consisted of a few part-time firefighters and an ancient pump.

In 1937 the Home Office instructed local authorities to draw up air raid precautions and fire protection schemes, and the Air Raid Precautions Act provided a grant to finance improvements in firefighting services.

These had grown up in a piecemeal way. Many firefighters were recruited from the navy, where they were taught the rigid and unquestioning obedience that fire officers believed the job demanded. Life in a fire station was run like life on board a naval ship, revolving around drill, discipline, spit and polish. New recruits were even expected to clean their officers' homes. "Fire stations," notes Victor Bailey in his history of the Fire Brigades Union, "were simply ships on dry land."

Every firefighter was on duty 24 hours a day for fourteen days and nights. Men who had fought a serious fire in the night had to spend the next day cleaning and testing the

engines and equipment. Discipline was arbitrary, and the station manager's word was law. Outside the big brigades, there were no pensions, and pay was reduced during sickness.

Few firefighters could rise through the ranks to become officers, for these were often recruited from the ranks of naval officers. The man who was to become easily the most famous of these was the appropriately named Commander Aylmer Firebrace, veteran of the 1916 Battle of Jutland and a former commander of Chatham Gunnery School. As we shall find, the instincts of a naval officer accustomed to instant obedience never quite left him.

Firebrace became an officer in the London Fire Brigade in 1919, and in 1933 he recruited to the ranks a young man of twenty-one who, arguably, was to have as big an influence on the future of the fire service as Firebrace himself. His name was John Horner, and he did so well at training school that Firebrace attached him to headquarters and placed him in the

Commander Aylmer Firebrace © London Fire Brigade

zone of accelerated promotion. If Firebrace had had any idea what sort of man young John Horner would turn out to be, it's unlikely he would have done that.

Horner found the LFB deeply conservative and rooted in the past, and was horrified to discover that its equipment was incompatible with equipment used elsewhere in the country. Could those in charge not see how dangerous this would be in time of war, when other brigades might need to come to

John Horner, Fire Brigades Union General Secretary 1934-1964, addresses the Trades Union Congress on 7 September 1962. © Press Association

London's aid? Could no one see that military-style uniforms and drilling to make men docile were not the preparation required for war work? Apparently not.

"It was the brass helmet which for me epitomised the hidebound traditions of the LFB," he wrote later.

> *Polished every day and after every fire, the crown might become dangerously thin by generations of elbow grease, since a helmet could have previously served more than one fireman throughout his entire service. Heavy and cumbersome, awkward in confined spaces, it provided little protection. Since the dawning of the age of electricity the high crest with its splendid embossed dragon had been a constant danger to the wearer.*

Born in November 1911, Horner knew what poverty did to people and to families, for he remembered his father, who arrived in London a thirteen-year-old orphan, looking for work: "He was to remain an illiterate navvy until, his strength failing, he simply died." Horner's parents brought up their four children in a tiny two-up, two-down house in Walthamstow. The defeat of the 1926 general strike when he was fourteen seems to have been the point at which he became a socialist.

A scholarship to a grammar school led to him becoming a trainee buyer for Harrods. It could have been the gateway to a comfortable life far removed from the privations of his parents, but everything in this restless man revolted against it, and after a year he left and went to sea. He travelled the world with the Merchant Navy and became a qualified second officer – but one who could not find a ship because of the economic crisis of the 1930s. So he applied for a job in the fire service.

As a seaman, he recalled, "I had seen the collapse of world capitalism. I had seen the unwanted wheat of the prairies of America mixed with tar to fuel dockside locomotives, while starving jobless immigrants wandered the quays and begged for scraps from the hogswill in our galley's shit-bucket." Now

1942: The Minister of Home Security, Herbert Morrison, at the top of a ladder during the demonstration of London's highest fire turntable. © Getty Images

he was seeing the dreadful poverty and starvation of the 1930s in Britain; and he had seen the 1929-31 British Labour government do nothing at all to change it. It was a bitter disillusion for reformers of Horner's generation. So he was quite ready to be what his colleagues called "a bolshie".

Horner became an activist in the Fire Brigades Union, whose power and influence were very limited. Two thousand of its members were employed by the London County Council, and the other thousand were spread thinly over provincial brigades. The LCC refused to have anything to do with it, and set up a tame "representative body" (RB) instead. The LCC under Herbert Morrison made a parade of refusing to deal with the FBU. Morrison also poured scorn on the claim that firefighters worked 72 hours, "announcing publicly that we actually slept at nights – on beds – with blankets supplied by the LCC." Contemporary London politicians have recently, as I write in 2010, made the same appalling discovery, and speak about it with the same manufactured outrage.

The military-style drill was designed to make the men docile and unlikely to get involved with the union. But firefighters relied on each other, and as Horner wrote later: "I came to realise that the strong bond of mutual reliance which characterised the job could be a powerful element in forging a special kind of trade union for a special kind of service."

These two men, the grandee who had been a naval officer and the young working-class former merchant seaman, were to be the key figures in the fight against the Blitz.

In 1938 Firebrace was appointed chief officer of the LFB. That was the year of the Fire Brigades Act, which required local authorities to maintain efficient firefighting services, and Firebrace took up his post just in time to implement it in the capital.

Before this Act, local authorities provided whatever they thought they ought to provide by way of fire services. No local authority except the LCC was under any legal obligation to provide a fire brigade at all. What they did provide was often very sketchy indeed. The heads of local fire brigades were often local worthies who had an interest in firefighting.

The Fire Brigades Act of 1938 made fire protection compulsory for every local authority. Britain was divided into twelve regions under the command of Chief Regional Fire Officers to co-ordinate resources, including manpower and appliances. But the Act did not provide extra money, and it did not reduce the fire brigades to a manageable number. When tested in war conditions, as we shall see, the new Act proved insufficient.

But it was a sign that the government and the press had belatedly woken up to the reality of aerial warfare, and were tending if anything to exaggerate its potential for destructiveness. A report by the Ministry of Health commissioned in spring 1939 calculated that during the first six months of aerial bombardment 600,000 people would be killed and 1,200,000 injured. This proved to be a considerable overestimate because it was based on faulty assumptions about the number of German bombers available and the average number of casualties caused by each bomb.

The Fire Brigades Act put the expansion of the fire service under the control of the Home Office, which issued a Memorandum on Emergency Fire Brigade Organisation. This pointed out that the assumption that large numbers of fires would not occur at the same time, upon which fire service organisation was based, was reasonable enough in peacetime but clearly wrong in the sort of war that was likely to be fought. For war, a huge and rapid expansion was needed, both of men and equipment.

Frederick Radford, the first historian of the Fire Brigades Union, noted just after the war that "the Union drew the nation's attention to its woefully inadequate fire defences. There was a criminal lack of equipment and uniforms. Fires were fought in civilian clothes in some towns. There were deficiencies in hoses, blankets and beds."

Trailer pumps were specified and manufactured in a hurry, and great quantities of hose purchased from Canada and the USA. And then, in March 1938, the LCC's air raid precautions committee set up what became known as the Auxiliary Fire Service (AFS) by deciding to recruit and train 28,000 auxiliaries for the London Fire Brigade, and to set up 360 additional fire stations. There were to be two types of auxiliary: unpaid part-timers, who did their normal jobs and worked as firefighters when they could and when they were needed; and those who, in the event of war, would give up their jobs and become full-time, paid firefighters. Women were recruited both for driving duties and to look after the telephones.

The introduction of women was not likely to appeal to the crusty naval commander now heading the London fire service. Before the war was done, Commander Firebrace was going to have to get used to many things that he would not have dreamt of in his worst nightmares, and women firefighters were but the first, though among the most disturbing. He had a glimpse of the misery that lay ahead when, in the summer of 1938, he had to discuss the matter with two leaders of the Women's Voluntary Service.

"I resolved never to let myself be placed in such a situation again," he wrote furiously to A. C. Dixon, head of the Fire Services Division at the Home Office.

> *I object to being bullied on my own quarter deck by a woman . . . After this I appointed Lady Makins as Woman District Officer at headquarters and, incidentally, as my personal bodyguard against any similar treatment with which I might be threatened . . . Admittedly the women's branch of the AFS is behind that of the men's – the men's is more important, and we know better how to deal with our own sex . . . The WVS does not know enough of our organisation really to be able to help and the tone of the criticism I have so far received has left much to be desired.*

He was not alone in failing to see a useful role for women (and before the war was out he was to change his mind). Their traditional role was to clean and cook, and in fire stations men did those jobs, as they did at sea. What else could women do – apart, that is, from going out and fighting fires, which Firebrace and many of the men he commanded would have considered unthinkable?

In the event they became the nerve centre of the operation, doing some of the most dangerous work, taking calls, directing fire engines to the right place, riding motorcycles carrying messages and driving cars carrying fire officers. And often, though officially discouraged from doing so, they found themselves playing an active part in firefighting. If they could see help was needed, they rushed to provide it, and the hard-pressed men did not turn them away. They had to learn about the fire service very fast as war came close, because those in

AFS women dispatch riders in training, 1940s © London Fire Brigade charge had been so slow to realise that they would be needed.

An AFS was formed in every county borough, borough and urban district, and there was also one in the LCC. Each AFS was headed by a Commandant. The services operated their own fire stations, each commanded by a Section Officer.

An extensive recruitment campaign, including a radio appeal and widespread advertising, produced a few recruits, though not as many as had been hoped. Recruiting became quicker after the Munich agreement in September 1938. Neville Chamberlain's pact with Hitler, which gave Hitler what he sought in Czechoslovakia, and which Chamberlain thought meant "peace in our time", seems to have been what convinced many people that it was the exact opposite: that war really was coming, and firefighters were going to be needed to deal with the results of air attacks.

Auxiliary firefighters came from every trade and profession: sailors and solicitors, bricklayers and fishmongers, milkmen and mechanics, hairdressers and car salesmen. Two of the new post-Munich recruits were Hubert and Joyce Idle, who were attached to a fire station in Wandsworth, in south London. Hubert was a research chemist, which was a reserved occupation so he could not be called up to fight with the armed services, and Joyce was a secretary. They both felt that, with war now likely, they should add part-time firefighting to their daily work. Hubert was to see things that gave him nightmares for the rest of his life.

Some experienced London firefighters were assigned to give AFS men a 60-hour training programme.

Outside London, too, Munich was a spur to recruitment. In the Somerset village of Combe Down lived First World War veteran Harry Patch, who had served at Passchendaele and was later wounded by a shell that killed three of his comrades – that same Harry Patch who outlived all his fellow soldiers and was the last living First World War veteran when he died in 2009, aged 111. In 1938, Patch was a fairly typical AFS recruit: just too old for military service, but anxious to do something for his community in wartime, and with experience that might be useful to a firefighter – he had been a plumber. His fire crew operated from a temporary fire station established in one of the vicarage's stables. The fire engine was kept at ground level, and the hayloft was turned into sleeping quarters for the crew.

The idea was that there would be six AFS sub-stations attached to each of the 59 London Fire Brigade pre-war main stations. They were to be sited in a variety of buildings, including requisitioned garages, filling stations and schools. They would supplement the regular fire brigades. There were to be an extra seven sub-stations in the docks where the Germans were thought most likely to concentrate their bombing. These would supplement the regular London Fire Brigade and the 66 smaller brigades in the Greater London region, grouped into five districts, each run by an Assistant Regional Fire Officer.

The government was by now rushing to provide the equipment the new firefighters would need. The familiar red fire engines used by the London Fire Brigade remained at their 59 parent fire stations. The AFS had new grey ones. They also had two-wheeled trailer pumps which could be towed behind an ordinary car or van. A few vans were bought for the purpose, but nothing like enough, and in the end the solution came in the form of the London taxicab. The idea of using them was put to Herbert Morrison by the Transport and General Workers Union in September 1938. Over 2,000 of them were hired, often along with their drivers, by the London Fire Service before war broke out a year later. Emergency water supplies were installed in many towns, and where a large water supply such as a river was available, pipes were laid to provide water for firefighting.

In the AFS, men and women had separate officer structures: people with what was called a suitable background – which meant people from the professional classes – were appointed to command them. This was not a good idea. As the auxiliaries started to do the same work as regular firefighters, their officers started to clash with those of the regular fire service, who knew a great deal more about firefighting than they did.

But the biggest problem was the relationship between the AFS and regular firefighters. The AFS men were often resented by the regulars. John Horner summed up the way many of his colleagues felt about it. By September 1938, he wrote, "there were 18,000 invaders occupying our stations." The AFS men were even asking to be allowed to drive to fires with the regulars, on their splendid, highly polished fire engines. "These splendid creations, the darlings of their drivers, were now violated by over-enthusiastic strangers – clerks and shop assistants – playing at firemen," writes Horner.

Many regulars were not at all keen on the idea of women going into the AFS – firefighting, they believed, was man's work. Pay, too, was an issue. AFS people were to be paid £3 a week for men, £2 for women. Most regular firefighters earned more (though this was not true everywhere – there were no national pay scales).

On the AFS side there was resentment about the matter of uniforms. The brigade had only been allocated enough to give each AFS man one complete set of firefighting clothes, but the regulars had a spare set, so that when they returned to the station after an incident, they could change into a dry uniform before going out again.

What was the FBU to do? Should it man the barricades and repel the intruders, or welcome them in and take up their grievances? For a long time it did nothing. But John Horner, now on the union's executive, was sure that could not continue. The union had 3,150 members – but there were to be thousands of AFS firefighters in London alone. "If the AFS was mobilised before the FBU had found a constructive relationship with it, the union would be swamped out of existence," he wrote later. Horner began to convince his colleagues that they could not afford to have the AFS outside the union, or in another organisation – the government might well organise some sort of non-union staff association, which would instantly be many times larger than the FBU.

For the FBU, the problem was complicated by their negotiations with the LCC – for which they had to reconstitute themselves as the "Representative Body," since Commander Firebrace and Herbert Morrison made a point of refusing to talk to the union. If the FBU leader sent a letter on FBU-headed notepaper, it would be returned with a polite request that it be typed on RB-headed paper before it could be dealt with.

The RB – which effectively meant the FBU – had demanded a 48-hour week. Morrison was offering 60 hours, and the offer was contingent on all sorts of conditions – most of them trivial, but designed to make firefighters feel like serfs. Chiefly they were disciplinary measures of the sort of which Commander Firebrace approved: no smoking in the watch room, that sort of thing.

Nonetheless, Morrison's offer constituted a victory of sorts, and the union's leader, Percy Kingdom, tried to get the men to accept it. He was not successful. They followed Horner's lead instead, and rejected it.

Kingdom, one of the many ex-seamen who became firefighters, was described five years later by someone who knew him as "dour, determined and blunt". Under his leadership the union had increased its membership and had fought bitterly with Morrison over firefighters' hours. He was not a man who took rejection of his deal lightly; he resigned, nominating as his successor his assistant secretary. But the members elected John Horner instead, both as the candidate of the left and as the standard-bearer of the policy of welcoming and recruiting the AFS.

The other full-time officials resigned. Horner started the job with no staff at all, not even a typist, though the executive did agree that he might appoint one, with the proviso that no woman should be employed.

Horner set about recruiting the AFS, despite the hostility of many of his own members and a solemn warning from TUC general secretary, Walter Citrine, that it would be foolish to risk the union's meagre funds, and that it should concentrate on providing a service to its regular members. "You are attempting the impossible," said Citrine. As Horner set about recruiting the AFS, torn-up membership cards with "traitor" and "sell-out" written on them began arriving at the union's small Clerkenwell office. And that was before the regulars realised the final infamy: that Horner wanted AFS people to have the same pay, and the same rights of representation in the union, as professional firefighters.

Hitler invaded Poland on 1 September 1939. Two days later, Prime Minister Neville Chamberlain announced a state of war with Germany. At once, all LFB people were recalled from leave, and premises were quickly requisitioned as AFS headquarters. The same day, Home Secretary Sir John Anderson made LFB chief Commander Firebrace the new London Regional Fire Officer. Firebrace was furious – he had wanted to keep his command of the LFB as well as be the overall head of the region's firefighting services. But the Home Secretary was adamant, and Firebrace was forced to let his deputy, Major Frank Jackson, take over the LFB.

And just as the Prime Minister finished speaking, the first air raid warning sounded over London.

Chapter 2

3 September 1939–July 1940
Watching and waiting
and polishing brass

It was a false alarm. There was no bombing raid at the end of Chamberlain's declaration of war. The siren was the result of a French aircraft arriving at Croydon aerodrome and being mistaken for a German one.

But the AFS was ready, with its 2,000 motorised and trailer pumps in London alone to add to the 120 red engines of the regular LFB. They expected to be in action at once, and they had been told hair-raising stories about the huge capacity of Luftwaffe aircraft to carry and drop incendiary bombs. The historian Eric Midwinter, then a seven-year-old boy living in Sale, outside Manchester, remembers that

> *on the day war was declared, all the barrage balloons went up round us and in Manchester, because everybody was expecting that the bombing was going to start almost immediately. And I have a very vivid recollection of my father putting his uniform on, this black serge AFS uniform, and I was quite upset, because I thought he was going off . . . and I can remember him holding me and giving me this reassuring cuddle, and this thick black serge against my cheek.*

The fire engines were there, and so were the firefighters, but much else was missing. Many of the stations had no cooking or heating facilities, and others had no beds – men had to sleep on bare boards, with primitive toilet arrangements. "The complete lack of care for the auxiliaries in many brigades was nothing short of criminal," wrote the FBU's AFS national officer, Peter Pain, who had been a lawyer and was to return to the law after the war. "In some, men were sent out to stand-by day and night without any stations being provided at all. In the vast majority of cases, stations had atrociously bad living conditions."

Vic Flint, billeted in an infant school in Whitechapel, made a pastime of catching the lice. "Mac" Young in Paddington found that his sub-station, a Great Western Railway recreation hall, wasn't ready, and he and his 24 AFS colleagues bedded down amid the straw with 200 restless GWR horses, as well as a lot of flies, with no food and no washing facilities.

Perhaps fortunately, the expected bombing raids did not materialise in those first few months of the war. Instead of fighting fires caused by the Luftwaffe, AFS recruits were put to work filling sandbags and doing other heavy work, and sorting out their living and sleeping arrangements and their hastily assembled equipment. The full-time AFS recruits could not be kept permanently on duty when there were no air raids, and shifts were reintroduced: 48 hours on duty, 24 hours off. In the long term, this was to cause much resentment – it amounted to 112 hours a week, which was too long for anyone to be asked to work – but in the short term, recruits kept coming. In London there was a flood of new recruits, who no longer had to undergo a medical examination – they just had to sign a form saying they were fit. Two thousand seven hundred men and women enrolled during the first three weeks of war.

It was a miserable life for them. The winter of 1939–40 was ferociously cold and anti-freeze was scarce, so every fifteen minutes someone had to go out in the chill night air from their often unheated makeshift fire stations and start the engines of the trailer pump units and the taxis that towed trailers so they did not freeze. With no air raids, banned from going to ordinary fires with the regular firefighters, often housed in miserable conditions, doing repetitive exercises day after day and being required to do endless spit and polish, frequently cleaning already perfectly clean brasswork, the AFS recruits were a prey to boredom. (So were people outside the fire service. They were beginning to talk derisively of "the Bore War".)

But sometimes there were alleviations. Stacey Simkins, a teenage messenger at Size Lane in the City, found that the station had

> *a large recreation room which was equipped with table-tennis table, full-sized snooker table, piano and umpteen chairs. So when the phony war was on, before it all started, there was a fellow there – he was AFS – named Pincher Martin. He was always the one – when the piano was playing, he was the singer. And his favourite song was "Sioux City Sue". Funny the things you remember.*

Just such a recreation room – table-tennis table, snooker table, piano – is seen in the sub-station in Humphrey Jennings's 1943 film *Fires Were Started* (see Chapter 11). Caged rabbits and a kitten also feature, another true-to-life detail, for many fire stations had pets. Derek Godfrey, an AFS firefighter who wrote a book about his experiences, *We Went to Blazes*, published in 1941, recalled a dog which, his owner claimed, knew how to distinguish British aircraft from German ones. Another sagacious dog at Enfield Fire Station would join the crew on the engine whenever they were called out, once even jumping from a window twenty feet above.

*

In early 1940, Major Jackson gave permission for AFS men to attend ordinary fires.

He had given a lot of thought to this, knowing it would cause resentment among the regular firefighters, and he had done something that might perhaps not have occurred to Commander Firebrace: he had decided to try to enlist the support of the FBU and its newly elected young firebrand of a leader, John Horner. He met Horner, guaranteed him support for his efforts to bring the AFS into the union, promised freedom of entry to all stations and facility time for union officials, and said he would make himself available to Horner at any time. They would work together to get regular firefighters to accept the AFS as equals.

Around London, most Chief Fire Officers followed Jackson's lead and gave Horner's organisers free access to fire stations. Outside London it was a different story. There were many areas where the authorities still seemed to equate trade unionism with treason. In Cardiff and Swansea Horner was tailed by plain-clothes police, who stood in his meetings and took shorthand notes of his speeches. In Birmingham an autocratic Chief Fire Officer refused to allow his men even to meet Horner (though Chief Fire Officer Tozer was to get his comeuppance before the war was over, as we shall see).

Horner was preparing his union for the war, and for a future as a much larger and more powerful organisation than it had ever been before. He saw that the arrival of the AFS represented not a threat but an opportunity for the union.

He also realised that Percy Kingdom's criticism that Horner lacked experience had some force, and he looked for, and found, a few older men who had been organising the union as long ago as the First World War, and who would take on key positions now. The government organised the country into fourteen regions to prepare for an invasion, and Horner had a regional office in each of them.

Horner, having been elected leader of a union of 3,500 members, found himself by mid-1940 leading one of 66,500 members, including 1,000 women. Most of the new members were AFS, of course, but the union was also overcoming opposition in the provinces and recruiting more professional firefighters.

*

In those quiet days, when there was not much for the AFS to do, a lot of people thought they were just avoiding being in the armed forces. They were called parasites, idlers, and "£3-a-week army dodgers", even though firefighting was not a reserved occupation: AFS men could be called up, and often were. They were contemptuously dismissed as a "darts club". The public attitude helped sap their morale.

The sense that no one cared about them was heightened with the passing of the Civil (Personal) Injuries Act, which said that an AFS man who was wounded or sick would be discharged after two weeks of incapacity. They would have to throw themselves on the mercy of the Unemployment Assistance Board, which would insist on a means test before helping them. On top of all this, there was continued tension between regular firefighters and the AFS.

So it's hardly surprising that many trained AFS men and women left, often volunteering to go to the front where at least their service was appreciated. Eventually the government moved to stop the exodus, rushing through a statutory order preventing full-time firefighters from resigning.

The regular firefighters had their grievances too. Under a pre-war agreement with the Fire Brigades Union they had been promised a shorter working week of 60 hours, but they were now on duty an average of 110 hours a week.

There was, however, time to learn new things. In these early months of the war, the Workers Educational Association started some pioneer work in civil defence headquarters and fire stations. In London and the south-east, organised discussion groups were starting. Many of them functioned entirely spontaneously, and without any outside help at all. Later the poet Stephen Spender spent much of the war helping to start and run them.

*

The fire service's first real involvement in the war came when Britain had to sweep her exhausted and defeated army off the beaches of Dunkirk as the German army closed in on them. Hitler's forces were advancing fast, having overrun Belgium, Holland, Luxembourg and France in little more than a couple of weeks, and Britain was in full retreat.

Vice Admiral Bertram Ramsay had had just under a week to prepare the biggest evacuation in history. By 26 May 1940, Ramsay had assembled 50 passenger ferries at Dover and 20 more at Southampton, together with an escort of destroyers, corvettes and minesweepers. They had to take roundabout routes across the channel to avoid mines and shelling. By the end of the first day, only two ferries had succeeded in berthing, only 7,500 troops had been rescued, and it was clearly impossible to use the port. So the naval ships were diverted to the beaches. But here, shallow water prevented the ships from getting closer than a mile from the shore, and smaller craft had to ferry the troops to the ships.

There was one alternative – a narrow concrete pier, never intended for ships, which they discovered they could use. But only one boat at a time could use it. So all available small boats were sent across the channel – pleasure craft, lifeboats, Thames sailing barges – often with volunteer crew who had never sailed out of sight of land before. A huge fleet of small boats arrived to get the British army (and a large part of the French army) off the beach.

The fire crews on the Thames, who had grown as tired as their land-based colleagues of waiting for something to happen, were at last to see war action. The newest and biggest Thames fireboat, the *Massey Shaw*, headed down the Thames, with a crew including both AFS and regular firefighters. She had never been to sea, except when she was delivered from the Isle of Wight, and she was built for firefighting, not seagoing. Now she was ordered to go to Dunkirk – first with the idea of fighting the fires started by German aircraft, but as she crossed the channel her mission was changed to one of scooping men off the beaches.

It was a dreadful night. After losing two dinghies in the attempt, the *Massey Shaw* took 65 men on board, which meant she was massively overloaded and the men were crammed in like sardines. She made her way like that back to Ramsgate overnight. The next day she returned to Dunkirk, ferried hundreds of men to troopships, and was the last vessel to leave her part of the beach, at 3.30am, arriving at Ramsgate four and a half hours later. On her next trip she rescued many of the crew of a French naval vessel that was sunk by a mine just 200 yards away from her.

It was the fire service's first taste of excitement and war service, but it was not enough to stem the growing boredom and disenchantment among both auxiliary and regular firefighters – especially among auxiliaries, because they still did not have many chances to fight ordinary, non-war-related fires, despite Major Jackson's decision that they could do so; they were called in to help only at the biggest fires. It cannot have helped that the country was every day expecting an invasion. Scrubbing brass and doing drill must have seemed a pretty pointless way to spend one's days.

A few Luftwaffe sorties were made in May and June, with bombs falling in open countryside in Kent and Surrey. Firefighters became a little busier, but it was still not the firestorm they had been promised. They were starting to wonder if that would ever come. There would not be long to wait.

Chapter 3

July–7 September 1940
From the Battle of Britain to the Blitz

Between 10 July and 11 August the Luftwaffe made its preparations for what Churchill dubbed the Battle of Britain, which happened in the sky during the summer and early autumn of 1940. German planes reconnoitred the British coast, and made a few preliminary daylight attacks on coastal towns including Wick, Hull and Dover. It was a battle for air superiority, and Hitler's intention was to pave the way for an invasion of Britain.

Between 12 and 23 August German bombers attacked coastal airfields. There were lightning raids on many southern RAF airfields, including several near London, at Biggin Hill, Croydon, Kenley and Hornchurch. In the words of an account published two years later by the Ministry of Information, *Front Line 1940–41: The Official Story of the Civil Defence of Britain*, "The aerodromes mattered most; the fires upon them numbered scores and hundreds, big and small, and to fight them under bombing became almost a matter of routine. At Manston [in north-east Kent], late in August, firemen fought fires in hangars and stores for two days and nights on end." There was a raid on Croydon airport on 15 August, but the first incendiary bombs to fall on the city of London did so on 17/18 August, in Woolwich and Eltham in the south-east of the capital.

From 24 August to 6 September there were concentrated attacks on all airfields, and on ports and aircraft factories. Fires were still being dealt with by local firefighting units – it was not until the Blitz itself that bombed areas started to call in reinforcements from outside. In this period the Luftwaffe also attacked fuel tank depots at Thameshaven on the Essex shore of the Thames, but failed to locate Thameshaven's flammable fuel tanks. So, during the raid on Thameshaven on 24 August, some German aircraft strayed over London and dropped bombs in the eastern and northern parts of the city: Bethnal Green, Hackney, Islington, Tottenham and Finchley. Air raid sirens sounded at 11.08pm as incendiary bombs clattered on to rooftops and burst into intense balls of fire. Some penetrated roofs, others lodged in guttering. Several buildings caught fire, including two dockside warehouses.

London Fire Service crews, regulars and auxiliaries together, were out within seconds, and saw at once that if they did not get the flames under control, the two warehouse fires had the potential to threaten all surrounding buildings. At Fore Street on the fringe of the City of London, where some of the fiercest fires raged, 200 pumps were used, mainly AFS heavy and trailer pumps but also some LFB appliances, each with a six-man crew. By mid-morning the next day the fires had been beaten, though fire crews stayed while they cooled down until well into the evening.

This prompted the British to mount a retaliatory raid on Berlin the next night with bombs falling in Kreuzberg and Wedding, causing ten deaths. Hitler was said to be furious, and on 5 September, at the urging of the Luftwaffe high command, he issued a directive "for disruptive attacks on the population and air defences of major British cities, including London, by day and night".

On the very day that Hitler gave his order, the Luftwaffe had another go at the fuel depots at Thameshaven, and this time they scored a direct hit, setting five 2,000-ton oil tanks ablaze. The fire was far too big for the local brigade to handle alone, and for the first time, outside reinforcements were sent for: they asked the LFB for forty pumps. Unfortunately, when they arrived, an AFS volunteer firefighter who was in charge said they were not needed, and sent them away again.

It took several hours, and several infuriating bureaucratic

delays, to get them back again. The volunteer firefighter who had sent the pumps away went home, and a London officer asked the Lambeth headquarters for the 40 pumps back. He was told that the rules said only a local officer could make the request, and a London officer must not usurp his authority. The London man explained that no local officer was really in charge any more, and Commander John Fordham, a senior LFB divisional officer, was sent to the fire to make an assessment.

Fordham was a forceful man with red hair and left-wing views, and did not suffer fools gladly. He saw at once that the pumps were desperately needed straight away, along with three fireboats. He too was told that local authorisation for the request was required, but he was not a man to take no for an answer, and threatened to ring everyone – the Home Office, even the Prime Minister.

Apparently, in the absence of a local officer, the order had to go through the regional commissioner for Essex and East Anglia, who was, it turned out, the Master of Corpus Christi College, Cambridge. Efforts were made to contact this eminent gentleman, and at last Fordham was told that the Master had retired for the night, and that his staff were reluctant to wake him.

Fordham exploded, and they promised to send someone to take a look. Eventually the Master's emissary turned up in a sports car. Fordham's description of the "young fellow in a sports jacket" who introduced himself as being "from Cambridge City Surveyor's Department" became part of fire service folklore. The young fellow stared, horrified, at the blaze, and confessed he knew next door to nothing about firefighting. He had, apparently, once attended a week-long firefighting course. He said that, as the representative of the sleeping Master of Corpus Christi College, he was authorising Commander Fordham to take control, and take what action he considered necessary. Fordham ordered up the pumps and fireboats at once.

Meanwhile firefighters were working up to their necks in hot oil to get the fire under control. James Gordon, a former journalist turned AFS man, was in one of them, and wrote afterwards:

The burning tanks flung up dense, black, oily smoke which looked almost solid and effectively obscured all signs of flame. Past tanks battered, squashed and melted into fantastic shapes. Past men unrecognisable in squelching veils of oil . . . They were all dismayed, even Tommy and Ginger. Their tunics were new, their boots and leggings well blacked. Now they were ordered to get to work on a burning tank which stood in a concrete pit which was deep with oil.

In the concrete pit, they were conscious of being inches from a dreadful death. If some of the burning oil fell into the pit, they would all be fried. Gordon takes us into the minds of one of the men: "He was a mug to have volunteered for this job. If anything happened to him, his wife and kid would suffer. He felt the oil creeping up his leggings and wondered what his wife would say about his ruined pants."

Fordham knew, and so did John Horner, that the several hours' delay had put the men in additional unnecessary danger. "I believe that the idea of a *national* fire service was born in Fordham's mind that night at Thameshaven," wrote Horner. "He became a propagandist for reform and between us we maintained an uneasy contingent alliance for years." But it was to be some time yet, and there were to be many more bureaucratic muddles over control, before Fordham and Horner got their national fire service.

The Luftwaffe were back the next night, to attend to some of the oil tanks they had missed before. The night after, the planes appeared in the sky above Thameshaven yet again. But this time, most of them flew straight past. This was 7 September, and they had bigger fish to fry.

Chapter 4

7 September 1940
The first night of the Blitz

What we know as the Blitz began in the early evening of Saturday, 7 September 1940, when 364 German bombers, escorted by 515 fighters, flew across the Channel and followed the Thames estuary to London. Despite blackout restrictions, the Luftwaffe had a relatively easy way of getting to their destination. They simply had to follow the route of the River Thames – which also directed them to the docks based in the city's East End. Each night Hurricanes and Spitfires hastily engaged them, and duelling aircraft wheeled and turned in the London skies.

London air raid sirens sounded just after 4.30pm as the first bombs came down on the East End and the docks and ammunition dumps around it. Just a few minutes later the bombers had penetrated far enough to attack the quays and warehouses of Surrey Commercial Docks, piled high with timber. Fires spread instantly from timber stack to timber stack, and quickly embraced a whole square mile. Hundreds of small terraced homes, as well as shops and factories, were either blown to pieces or caught fire.

In the East India Docks, thousands of gallons of rum were blazing in one warehouse, gushing liquid fire across the surface of the water, as though it were a giant Christmas pudding and someone had poured on the rum and put a match to it.

The Germans continued bombing for 90 minutes. Anti-aircraft fire was sparse and ineffective. When the "all clear" sounded at 6.15pm, wrote an onlooker five miles west in Soho, "the whole sky to the east was blazing red . . . it seemed as though half of London must be burning, and fifty thousand

firemen would not be able to put out a fire of that size. In Shaftesbury Avenue . . . it was possible to read the evening paper." The young AFS messenger Stacey Simkins watched the flames "making the sky look like dawn was breaking in the middle of the night." The FBU's general secretary, John Horner, and its president, Gus Odlin, stood at the top of Chingford Mount and saw London stretched below them,

dark now with the searchlights dimmed as the bombers flew homeward. A few miles to the east of us the very firmament was alive with flame and smoke. The loom of the fires in the docks was seen at Bedford and beyond. Helpless and frustrated, we stood together, witnessing the first terrible round of a battle which our members were to fight without respite for the next 57 nights.

About four out of five of London's AFS men had never before seen a real fire. But their dispatch riders rode through the bombs and the smoke, from fire to fire, and reported back to local controls. Sometimes Boy Scouts guided fire engines to where they were most needed, becoming known as the Blitz Scouts. As bombs cut many of the telephone links, these messengers were the only means of ordering men and appliances to the appropriate places.

These young messengers, teenagers mostly and many of them still at school – the minimum age was sixteen – rode their bicycles through streets lined with burning buildings, while enemy aircraft droned menacingly overhead, carrying crucial messages. When the need arose – though they were not supposed to do it – they would pile in with the firefighters and help to hold the hose. One of them, Eric Applebee, now 86, says, "You grabbed every opportunity you had to do something useful." He held a hose on that first night, climbing up a ladder and helping to save the clock tower on top of a brewery. "If I'd been caught," he says, "I might have been discharged, and then again I might have got a medal."

Fire at Surrey Docks during the Blitz, 1940 © London Fire Brigade

Firefighters aimed powerful jets of water at the fires, and in the docks they put pump suctions down in the deep water. After months of near idleness, the huge reserves of equipment and auxiliary firefighters went into immediate action, and Major Jackson knew he could have done with twice the number. As they worked, bombs fell all around them. They would duck, then carry on.

In the docks, everything seemed to be ablaze; there was fire in every direction they looked. Nine fires were out of control at 6.30pm, half an hour after the bombers had retreated down the Thames.

At times, fighting the fires seemed to be useless. In Surrey Docks, water in the burning wood stacks would be dried by the flames within seconds. The crew of a fireboat, trying to get close enough to have some effect on the flames there, found all its paintwork scorched off in seconds. Dockside cranes weakened and sagged under the intense heat, finally keeling over amid a huge shower of sparks.

Each warehouse seemed to have its own kind of fire, recalled one firefighter:

There were pepper fires, loading the surrounding air heavily with stinging particles so that when the firemen took a deep breath it felt like breathing fire itself; . . . a paint fire, another cascade of white-hot flame, coating the pump with varnish that could not be cleaned for weeks. A rubber fire gave forth black clouds of smoke so asphyxiating that it could only be fought from a distance, and was always threatening to choke the attackers.

Sugar, it seems, burns well in liquid form as it floats on the water in dockland basins. Tea makes a blaze that is "sweet, sickly, and very intense," It struck one man as a quaint reversal of the fixed order of things to be pouring cold water on to hot tea-leaves . . . A grain warehouse on fire brings forth unexpected offspring – banks of black flies, rats in hundreds, and as the residue of burnt wheat, "a sticky mess that pulls your boots off."

At the Royal Arsenal in Woolwich, ammunition dumps were blazing, and firefighters faced the extra danger of exploding projectiles flying through the air. Many of the water mains were damaged, and many fire hydrants dry, but fireboats were able to pump up to four tons of water a minute from the Thames and pipe it to those ashore.

AFS man Frank Hurd was at Beckton Gasworks and wrote:

Gasometers were punctured, and were blazing away, a power house had been struck, rendering useless the hydraulic hydrant supply – the only source of water there. And then overhead we heard Jerry. The searchlights were searching the air in a vain attempt to locate him. Guns started firing, and then I had my first experience of a bomb explosion. A weird whistling sound and I ducked beside the pump. Then a weird flash of flame, a column of earth and debris flying into the air, and the ground heaved. I was thrown violently against the side of the appliance.

It was important to maintain contact between the units fighting the fires, the city control centre and the regional control centre, when the telephone lines were down. These young dispatch riders carried messages requesting more equipment or personnel. Photo courtesy of exetermemories.co.uk

Reaching an alleyway just twenty feet wide, Hurd saw a fractured and burning gas main.

> *The flames issued from wide cracks in the road surface. Gas main fires must not be put out, as this would leave escaping gas with great risk of explosion. We were working with our backs against one side of the street, aiming our jets into the warehouse, with the street alight in front of us. The top floor of the building against which we had our backs kept bursting into flame.*

Two hours after the first wave of bombers, at 8.30pm, came the second wave, which used the still raging fires as markers. This was to be a pattern of the bombing – the first bombs dropped were incendiary bombs designed to give the following bombers the most obvious of markers. Aylmer Firebrace recalled the effect of being in the midst of an incendiary shower:

> *One moment the street would be dark, the next it would be illuminated by a hundred sizzling blueish-white flames. They made a curious plop-plopping sound as they fell on roads and pavements, but this was not often heard above the shrill whirring noise made by the pumps. They never gave me the impression that they had been dropped from the skies – they seemed rather to have sprouted.*

And after the incendiary bombs came the high explosives.

Firefighters were in the path of the bombs again, obliged to stand and point jets of water at the flames while bombs and burning shrapnel fell all round them, their spines shaken by the impact while they worked. The official contemporary account of wartime civil defence, *Front Line*, recorded in 1942:

> *The fireman . . . worked at times in heat that blistered the paint on the pump, and turned to steam the water of his jet before it reached its mark. He was taught to "get at it," to close in, with*

Fire at Surrey Commercial Docks, 1940 © London Fire Brigade

his head held down near the branch where the jet's draught made a channel through the smoke, inching forward until he could pour water on the fire's red roots. This might mean perching a ladder on the steep slope of a roof and climbing fifty feet. It might mean taking a quick chance under burning rafters, or in a corridor roofed with stone cracking in the heat. Having got into a building, he might find himself lost in utter darkness, unable even to find his hose and trace his way back.

He saw the broken bodies of comrades tossed high in the air with their pump by the direct hit of a bomb. He saw walls fall on them, roofs crash through buildings where they were at work . . . Many a time he saw a building that he had quenched by hours of toil re-ignited in an instant by the fall of another bomb.

Roy Sargeant, then a young ARP messenger for Fulham Civil Defence Council, remembers:

We of the Civil Defence had the greatest respect for the NFS. While we, naturally enough, would be burrowing down low when the bombs were dropping, the man aloft on the turntable ladder would be exposed to all the effects of blast and shrapnel – and staying there! Others, forced into taking long shots by the ferocity and extent of fires, would often work from vantage points within range of tottering walls and falling debris, all with a sardonical courage of taking not the rough with the smooth but the rough with the harsh, as a matter of Brigade routine duty.

Four of the auxiliary fire stations within the dockland area had to be abandoned, and were engulfed in the flames. At a regular LFB fire station, the women still working there had to leave their telephones to tackle small fires that threatened the building – though all the equipment was out with the men, tackling the huge fires. The women also ran a dressing station for firefighters brought in with burns and cuts, or blinded by the smoke.

Fires raged through the night, on both sides of the Thames, with a continuous 1,000-yard wall of flame below Tower Bridge. Moored boats were alight, and huge barges, their moorings burned, floated out of control down the river, burning as they went.

It was not until 3am that firefighters began to gain some sort of control over the fires. They had been working since 5pm the previous day. Their work would not be finished for another two hours, but at least they could pause and the lucky ones got some hot tea and a sandwich from canteen lorries staffed by women firefighters. By the time the "all clear" sounded, at 5am, they had worked non-stop for twelve hours.

In that time, 430 men, women and children were killed, including seven firefighters, and many more firefighters were injured. More than 60 boats had been sunk, and the docks were destroyed, with many buildings still burning. Work to subdue

Miles of hose was used to fight the big fire at St Katherine's Docks which blazed for over 12 hours on 9 August 1940. The London Food Depot was heavily damaged.
© Robert Hunt Library / Mary Evans

the remaining fires went on. Major Jackson announced that all personnel were to remain on continuous duty.

After the months of enforced idleness, the firefighters were profoundly dismayed by this experience. A man who returned from leave the next day found all his colleagues in shock, convinced they would not live for more than another week. Men who were old enough to have fought in the First World War said that the western front offered nothing worse than they had seen that first night of the Blitz.

AFS firefighters taking a tea break, 1940 © London Fire Brigade

Chapter 5

8 September–3 November 1940
Fifty-seven days of non-stop bombing

Bombing was sustained from 7 September 1940 to 10 May 1941. In one spell, London was bombed for 57 consecutive nights. During this first phase of the Blitz, raids took place day and night. Between 100 and 200 bombers attacked London every night but one between mid-September and mid-November.

Britain was not equipped to fight the bombing. Few anti-aircraft guns had fire-control systems, and the underpowered searchlights were usually ineffective against aircraft at altitudes above 12,000 feet (3,600m). Even the fortified Cabinet War Rooms, the secret underground bunker hidden under the Treasury to house the government during the war, were vulnerable to a direct hit. Few fighter aircraft were able to operate at night, and ground-based radar was limited.

During the first raid, only 92 guns were available to defend London, though within five days there were twice as many, with orders to fire at will. This produced a much more visually impressive barrage that boosted civilian morale and encouraged bomber crews to drop before they were over their target, though it had little physical effect on the raiders. It boosted firefighters' spirits, too. One of them told Stephen Spender that the guns "break the Jerries in half in mid-air". It wasn't true, or at least it wasn't often true, but it was wonderful for morale.

Londoners became familiar with the eerie palette of the night sky during an air raid, here described by Roy Sargeant:

a brilliant firework display of mixed yellow searchlights, our anti-aircraft shells brightly exploding, sending their showers of shrapnel like sparks. Barrage balloons (to discourage dive bombing) were outlined and occasionally came down, well alight, and all against a wide horizon of bright flashes, varying hues of red and yellow with clouds of multicoloured smoke. Horribly beautiful.

And he adds: "Since those days, public firework displays are of minimal interest to me."

An AFS firefighter who fought the Rum Wharf blaze in the East India Docks noted that

Occasionally we would glance up and then we would see a strange sight. For a flock of pigeons kept circling round overhead almost all night. They seemed lost, as if they couldn't understand the unnatural dawn. It looked like sunrise all around us. The pigeons seemed white in the glare, birds of peace making a strange contrast with the scene below.

During 1941 and into 1942, the journalist G. W. Stonier, assistant literary editor of the *New Statesman and Nation*, wrote under the pseudonym "Fanfarlo" a series of humorous vignettes, "Shaving Through the Blitz", that were published monthly in *Penguin New Writing*. In March 1941 he recalled the early days of the blackout, before the bombs started to fall:

the carefree evenings when we all used to go for strolls in the new-found dark . . . "Sandbags!" we would exclaim as we picked ourselves up and went on to discover lamp-posts . . . cigarettes flicked, a dialogue on street corners . . . gaiety had lost nothing with the lights down … a gross amiability, the adolescent pleasure of being heard, but not seen, infected everyone who was being nudged, shoved, swept along and held back by the stream.

But "How it has changed in the last eighteen months!" In the Blitz,

All that has disappeared – the lingering, the voices, the cigarette dream; and now with darkness falls the hush. Emptiness, but with every cranny filled. London has been given over to a monstrous drama, an act of darkness from which ordinary people, you and I as individuals, shut ourselves away. Earth and sky contract to form the arena; the city puts up its searchlights, a beetle laid on its back and helplessly waving its legs, while the hornet drones overhead; night after night the assailant returns, the victim quivers with upturned belly.

*

After 7 September, and once they had got over the shock of that first dreadful night, the spirit of the AFS men shot up. At last they were doing useful work, and everyone could see it – in fact they were doing the most important thing there was to do. Instead of being sneered at as army-dodgers, they were

---

Attaching a hose to the nearest pavement hydrant. Everyone worked together – there's a firefighter, an ARP man and a civilian volunteer in this picture.
© Robert Hunt Library / Mary Evans

cheered as they went home. There was no more waiting, and no more boredom. Boredom was the one thing no AFS man could now suffer from.

Stacey Simkins was a fifteen-year-old AFS messenger. Having worked in a similar job for a tea broker, he was familiar with the geography of the City, its side streets, alleys and courtyards. Seventy years later he remembered his first experience of bombs falling around him.

> *We were more* interested *than anything else . . . in what it was like. You didn't realise so much what would happen if one of them [bombs] dropped near to you . . . You'd be more interested in what you were* doing *rather than the possible effects. I found this right through the whole war . . . I was more intent on doing what I had to do rather than thinking, "What happens if . . . ?"*

A London Fire Brigade vehicle in a bomb crater after an air raid in West India Dock Road, 1940. © London Fire Brigade

The journalist Negley Farson, writing in 1941, made a similar observation.

At the beginning . . . the average Englishman was far too busy to think very much about what he would do about anything . . . The A.F.S. people, the air-raid wardens, the police, the women ambulance drivers – all the great mass of auxiliary and regular city services – carried on because that was their job . . . It was automatic . . . its individual, instinctive heroism . . . was not a premeditated act. I think it is true to say that in those first startling nights very few of the people involved in the actual fire-fighting or rescue work made any conscious call upon their courage. They just met it.

Roy Sargeant, a messenger like Stacey Simkins, but with the ARP, remembers the feeling of being simultaneously fascinated and fearless.

One day my dad directed me not to go out with the wardens on patrol during air raids but to stay in the shelter. He told me that when he came home after his bus conductor work (a tough job in the blackout and raids), he would never go out in an air raid but would stay in our personal street shelter all night long.

At fourteen I knew, without doubt, that I had a charmed life. Impossible for me to be killed. It took a while for me to understand that I was just as vulnerable as a normal, even ordinary, common or garden mortal!

However, that night, in flagrant disobedience, I was on patrol with two wardens when a string of incendiary bombs scattered along Fulham Road. Hearing something coming down, we knew not what, we threw ourselves backwards into the entrance of a dental surgery and I collided with a civilian behind me. Turning to apologise, I found it was my dad – not in the shelter but out in the street to deal with the incendiaries. Not even a steel helmet! He looked, pointed his finger straight at me. I had been sprung –

but so had he.

My dad had been a light infantryman in the First World War and had fought in many of the major conflicts. Four wounds, shrapnelled, bayoneted and shot. He understood. The incident was never mentioned between us – and he chose never again to tell me not to go on patrol in the open.

AFS man Hubert Idle remembered all his life his own fire station in Wandsworth being hit, and some of his friends being killed. But he also wrote later, in a memoir kept now by his son Angus Idle, "The great reward is the gratitude of people when their homes are saved. It is good to see the people return to their homes even though you are wet through."

He also remembered being tired all the time, with a kind of all-consuming tiredness. One night, when he had been driving a taxi with a pump trailer all over London, down strange streets with only small side-lights to see by, he ended the night at a coffin-making factory. It had been a cold, damp night, but

it was a lovely morning, about seven o'clock, the water supply was steady, the pump was running smoothly, no help was needed on the branch, in an hour with any luck I would be driving the crew home across London. I climbed into the driving seat to rest for a few minutes, but unfortunately I shut my eyes and they found me five minutes later asleep.

He was a man who could see the funny side of most situations. Called to a fire at a church, he wrote that he saw the monks standing outside watching the flames. Later he thought that was odd: it was a Methodist church. On enquiry, it turned out that the "monks" were men from the Gas Board in their asbestos frocks.

"One gets rather callous on the job," he wrote. "At a completely burnt out building, a chimney stood in solitary but

rather dangerous splendour. A challenge from our firemen –
'Bet you can't knock that pot off' – 'Betcha we can.' Up went
the hose, down came the pot."

Night two of the Blitz saw 44-year-old station officer Gerry
Knight fighting yet another fire. "It is said you never hear the
one that's got your number on it," writes firefighter and
historian Cyril Demarne. "Rather difficult to prove, I'd say. So
we'll never know if Gerry and Auxiliary Fireman Dick Martin,
who was working near him, heard it. In those days, LFB station
officers were issued with thigh-length boots. Gerry Knight's
served to identify his few remains."

Just because death is routine doesn't make it easier to deal
with. Betty Barrett overheard a sub officer telling the station
officer, "Gerry Knight has had it; stopped one almost to
himself." She was desperately shocked. "He was a lovely man,
so kind to us girls. I broke my heart over the news and cried for

*The aftermath of an air raid on Piccadilly, central London; firefighters hosing the
smouldering remains of damaged buildings, 15 October 1940.*
© Mary Evans Picture Library

days. He used to smoke Gold Flake and always gave us a cigarette. He knew we only earned £2 a week and I kept remembering those little kindnesses."

By the next night, Firewoman Barrett had her own troubles. Her parents were bombed out, and at first she thought they were in the house at the time; but she found them in separate rest centres.

*

After 7 September, fire chiefs knew they had to clear up the administrative muddle and infrastructural shortcomings very quickly. There was confusion about who had authority to order what, the chain of command at some major fires had not been clear, communications had not always worked effectively, and water had been short after damage was caused to the mains supply. As part of the attempt to clear up the muddle,

Charles Cunnington (front row, second from left) with other Oundle firefighters outside their fire station. © John Cunnington

promoted AFS men were given equal status with their counterparts among the regulars. This, said many regular firefighters, meant they could end up being commanded by "amateurs".

It was becoming plain to the firefighters that their work practices were going to be different from those of peacetime. They could not treat fires as separate incidents – the regional headquarters would have to monitor the whole fire situation in London and surrounding areas. The close-range firefighting they had known, with firemen working their way deep inside a smoke-filled building and avoiding using too much water because of the damage it caused, were over. The Blitz fires needed plenty of water, sent in by powerful jets.

Decisions had to be made quickly, because everyone knew that more was coming. The Luftwaffe was now in possession of French coastal airfields, from where raids on Britain could be launched very quickly.

One key decision was made during the day on 7 September: to call in reinforcements from outside the capital next time. The very next night, about 8pm, air raid warnings sounded and within fifteen minutes all the equipment and the men were out again, leaving the firewomen in the watchrooms co-ordinating the messengers. Huge fires raged all over London. The 200 bombers started a major fire near St Paul's Cathedral, though without threatening the Cathedral itself.

And this time reinforcements were sent from well outside London – Birmingham, East Anglia and the West Country all received messages from Commander Firebrace at the Home Office, and sent contingents, which arrived during the night and took over from some of the exhausted London men. Nearly twenty senior officers went from fire to fire in staff cars driven by firewomen, trying to co-ordinate the firefighting. Firefighters were the only people to be seen – apart from them, London's streets were deserted. Even police officers had instructions to take cover.

John Cunnington remembers his father, Charles Cunnington, a firefighter based in Oundle, going down to London. "They asked for volunteers, and he said, 'Yeah, I'll take a crew down there.'" They drove down in a grey AFS tender towing a trailer pump. "He came back looking absolutely shattered . . . filthy . . . about ten days later. And he never talked about it. But he used to say, 'It was *bloody awful.*'"

*

And so it went on, night after night. The first bombs would fall at about 8pm and fire crews would be out within minutes, rushing through otherwise deserted streets to try to contain the flames. Second and subsequent waves of bombers simply aimed at the fires they could see from the air, and at the firefighters who they knew must be there, and they went on bombing until about 5am.

An oil bomb came through the roof of the House of Lords on 12 September, but only one room was damaged, thanks to prompt action by the Parliament-based firefighters. Four days later, and just over a week into the Blitz, on 16 September, Major Jackson added up the firefighters' casualties so far for a report to the LCC Civil Defence Committee. The fire service had already lost two officers, nineteen firefighters and one woman auxiliary. Eleven auxiliaries were missing, presumed dead. Thirty-one regular LFB men, 120 male auxiliaries, three women and one youth were seriously injured. Three London fire stations had been hit, in Whitechapel, Southwark and Euston Road, and eighteen sub-stations were so badly damaged that they could not be used. Firefighting had become one of the most high-risk of all wartime occupations.

The day after Major Jackson presented those figures, auxiliary firefighter Harry Errington, a master tailor by pre-war trade, was injured and trapped in his fire station. Just before midnight on 17 September 1940, he and other AFS men were in the basement of a three-storey garage in Soho used as

a private air raid shelter and rest area for the fire services. A bomb struck and all three floors collapsed, killing twenty people, including six firefighters.

Thrown across the basement by the blast, Harry came to his senses to find fierce fires around him. As he dashed for the emergency exit, he heard the cries of a fellow firefighter trapped beneath debris. Harry could have kept going and summoned help when he got outside, but he didn't; he turned round and went back.

Wrapping a blanket around his head and shoulders for protection against the heat and flames, Harry started to dig the injured man out with his bare hands. He then dragged him up the narrow staircase, out into a courtyard and to an adjoining street. Despite bad burns to his hands (which caused him concern because they threatened his livelihood), he went back down into the basement to rescue a second man pinned

Fire appliances damaged in a bombing raid, 1940s © London Fire Brigade

against a wall by a heavy radiator, and carried him to safety.

Harry Errington was one of many heroes of the time, though one of only three wartime firefighters to receive the George Cross. But what he did not get was decent treatment from his employers. The Civil Injuries Act applying to the fire service said that injured men could be kept on full pay for only thirteen weeks. Errington's burns having failed to heal in the prescribed period, he was discharged from the AFS. "He was sacked on the same day that he received his citation from Buckingham Palace and a letter from the King," wrote John Horner in disgust.

In fact, even thirteen weeks was an advance on what had gone before. Until shortly before Errington was injured, AFS men got no sick pay at all, and were being dismissed, sent to the Unemployment Assistance Board and means-tested. You could find in the same hospital a regular firefighter on indefinite full pay and an AFS man sacked after a fortnight. It had taken a strident campaign orchestrated by Horner to get even the thirteen weeks Errington had.

On 18 September 1940, Major Jackson relaxed the continuous duty system, so that firefighters had 48 hours on duty and 24 hours off. But there was no other leave. The long hours added to the firefighters' hardships, and it required all their commitment, both to their job and to winning the war, to keep them doing it. It did not help that the Home Office refused officially to meet their representatives. Jackson, who saw the foolishness of this, pulled all the strings at his disposal to arrange an informal meeting between John Horner and Sir Arthur Dixon, head of the Fire Services Division at the Home Office.

Horner wrote of that meeting with Dixon:

I said that we had four mobile fire service canteens for the whole of London. Men cut off from help had been known to drink water from the Thames. We had . . . no emergency rations; the few cooks we had in the stations were at home during the night. Men came back to their makeshift depots, sodden, tired out and filthy, with no means of drying their only uniform and liable to be called out again.

True, a week or two earlier the government had relieved injured men in hospital of their liability to contribute towards the cost of their medical treatment, but I told him that discharging injured men from the service after three weeks, and sending them to seek help from the assistance board, was doing more to destroy the morale of the AFS than any bombing.

Jackson believed the Fire Service could not have got through the war without the union's help and co-operation. "When [Herbert] Morrison retired him from active service, he received no honours, no knighthood, but 'Gentleman Jackson' deserves honourable mention in this history," wrote Horner later.

The heaviest attack of the war so far – by 400 bombers and lasting six hours – hit London on 15 October. The RAF opposed them with 41 fighters but shot down only one Heinkel bomber.

By 3 November, the Germans had dropped more than 13,000 tons of high explosive and more than a million incendiary bombs for a combat loss of less than 1 per cent (although some aircraft were lost in accidents). The bombs were eighteen-inch cylinders with an impact fuse, so that they exploded and burned furiously when they hit anything. There were also oil bombs, about the size of a dustbin, which caught fire on impact, and sent out burning oil and metal splinters. Sea-mines, weighing a ton or a ton and a half each, floated down on parachutes, also exploding on impact; but some lodged in the soft mud of the Thames, where they did not detonate until they were disturbed, perhaps by a fireboat hastily putting out to douse other fires.

There were nothing like enough public bomb shelters. The government feared that, if people were provided with large, central deep shelters, it would create a "deep-shelter mentality" – people would descend into the shelter and, feeling secure there, would refuse to come out. A large, crowded shelter would also be an ideal breeding ground, not only for disease but also for rumours and panics. At first Londoners were not allowed to use the tube stations for shelter, but this caused an outcry and eventually the authorities relented, making use of about 80 Underground stations to shelter up to 177,000 people.

Those who spent the night in a tube station might have been relatively safe (though several stations – Balham, Bank, Bounds Green and Marble Arch – were hit and dozens of lives lost), but they were not notably comfortable. People had at first to sleep on stone platforms and contend with wind gusting through the tunnels. Later makeshift bunks were provided. Yet queues would form outside the stations in the morning, so that the first to get in when they were allowed to, at 4pm, would be able to find a place.

Another frequent response to bombing was what became known as "trekking". Tens of thousands of civilians slept far from their homes – in parked cars, taxis and buses; in churches and barns; even out in the open, on Hampstead Heath or in Greenwich Park – and walked, cycled or took buses into work and out again every day.

"Others," according to the historian Constantine FitzGibbon, "went even further afield. For nearly two months several hundred 'unauthorised evacuees' were living in the Majestic Cinema in Oxford, eating, sleeping and, to the disgust of the burghers of that ancient and unbombed centre of learning, even copulating among the cinema seats."

There was little defence, and inadequate shelter. The major successes in the Blitz were scored by the fire service. Their biggest problem was getting enough water. Even when the water mains were working, the demands made on them were greater than they could supply; and they were easily damaged by the bombs. Long lengths of hose pumped water great

distances, but required great skill to keep in working order.

The consequences of water failure could be tragic. Charles Harbud was an AFS firefighter based at Parnell Road Fire Station in Bow. In 1946, he took his son Frank to the site of one of his wartime call-outs.

> *A bare, rubble-strewn, bombed London landscape stretched before me. My father was standing on a cracked and fragmented pavement, staring into the distance.*
>
> *"It's here," he said, pointing with the toe of his shoe at a wavy line at his feet. On close examination, it proved to be the remains of a corrugated iron sheet that gave the appearance of being burned off crudely close to the ground.*
>
> *"What was here, Dad?"*
>
> *"One of the largest wood yards in the London docks – or it was until it went up in smoke. The heat turned this corrugated*

Air raid shelter, Holborn Station, 30 January 1940. The London Passenger Transport Board had by then fitted station platforms with bunks, enabling Londoners who used the tubes as shelters to get to sleep. © Science & Society Picture Library

iron fence red hot, twisting it and melting it. We couldn't get anywhere near it for days, and we were damping down for a month."

Gradually I drew the story from my father. A crew with their appliance had been directed down the road that divided the yard in two, in order to prevent the flames from the half of the yard that was alight spreading to the other side. Unfortunately, either sparks or further incendiary bombs set alight the wood between them and the exit. The trapped crew were still able to fight the fire. Then a bomb fell and blew a crane over and blocked the road between them and escape. The crews on the outside kept the water relay to the inside crew going and were in fact winning the fight against the flames and the battle to extricate the crew from their predicament.

"And then the bloody tide went out." The anguish in my father's voice I can remember to this day. "There was no water! The heat, noise, wind, as the fire ran away. Thank God we could not hear them.

"Out there, in amongst millions of cubic feet of wood ash, are the ashes of six men. We never found them. To be quite honest, we didn't want to look too hard. We buried sandbags, not men."

John Horner happened to be there the night they bombed Soho Fire Station, where Harry Errington had been based before his injury. Horner was driving in the West End, and he knew from the noise of the explosion that a bomb had fallen a few hundred yards away. He drove into Shaftesbury Avenue and saw a huge cloud of smoke and dust. Walking through it, he found that a bomb had sliced off the top floor of the fire station, and the ceiling of the appliance room had collapsed.

"Already the men who had survived were digging into heaped debris . . . Girls in the control room were badly injured and three LFB men who had been in the mess were dead." He helped his members to search for their colleagues.

Chapter 6
3 November–28 December 1940
Turning on the provincial cities

As suddenly as it had started, on 3 November 1940 the bombing stopped. After 57 nights of continuous action, London's firefighters had an undisturbed night's sleep. But it soon became clear that all that had happened was that the Germans had shifted their target, and were now going for provincial cities.

Some had already suffered bombing. Birmingham and Bristol were attacked on 15 October. Now other important military and industrial centres, such as Belfast, Cardiff, Clydebank, Coventry, Exeter, Greenock, Sheffield, Swansea, Liverpool, Hull, Manchester, Portsmouth, Plymouth, Nottingham, Brighton, Eastbourne, Sunderland and Southampton, were to experience heavy air raids and high numbers of casualties.

Coventry, which was packed with arms factories, suffered a devastating attack on the night of 14/15 November 1940 which destroyed the centre of the city. It lasted for eleven hours, during which the Germans dropped over 500 tonnes of high explosive and 30,000 incendiaries, killing 554 people and injuring 1,200 others, and damaging 70,000 buildings. Within fifteen minutes of the first bomb falling on the city, the roof of the ancient Cathedral of St Michael was ablaze. Fires spread swiftly through Coventry's many old and closely built timber-framed buildings. By 8pm – less than an hour after the first bombs fell – the Central Fire Station had recorded 240 fires. Then the station itself was hit, and after that accurate record-keeping became impossible.

The city's own fire service fought the blaze themselves for four hours before reinforcements were sent for. Dozens of fire engines from nearby towns like Solihull dashed to Coventry to lend their assistance, but by the time they arrived the city centre was a great mass of fire.

In the Cathedral, firefighters were hindered by the construction of the roof: the inner wooden vaulted ceiling was separated from the wood- and lead-sheeted outer roof by an eighteen-inch space, within which incendiaries lay and blazed away, out of reach of the firefighters. Even when the Solihull Fire Brigade got there, their hoses were soon damaged and the water supply dried up. As water mains throughout the city were broken, it became obvious that there was little hope of fighting back the flames, and eventually the firefighters had to be content with removing whatever items of value they could and retiring to safety. By the end of the night the cathedral was in ruins. Twenty-six firefighters had been killed and 200 were injured.

John Horner drove into Coventry first thing in the morning, having stood during the night in a street in Oxfordshire and watched and listened as waves of Coventry-bound bombers swept overhead. Coventry FBU branch secretary George Dipper showed him the devastated city. They went together to a service for the city's dead, in a safe corner of the ruined cathedral. Horner retained all his life a clear memory of George Dipper, in his filthy uniform, walking up towards the ruin of the cathedral's three spires.

"It was an experience without precedent in the history of any British city," according to *Front Line*, " – a terrible test of the strength of the defensive machine. Inevitably there were failures and weaknesses . . ." But since this was an official publication, these failure and weaknesses were not detailed. The truth was that the firefighting effort had not been properly co-ordinated and there was nothing like enough water. For some reason, someone decided to send 50 AFS men to Coventry on buses, without their pumps. It did not matter that much; the order had come too late for them to be able to do a

great deal, even if they had been properly equipped. About the only gleam of light was that the brewery had somehow escaped, and the firefighters had as much free beer as they wanted.

Lessons were learned. In the following week the Luftwaffe twice attacked Birmingham, and this time the LFB quickly sent fifteen pumps and two water units and crews to help the Birmingham brigade.

Ruins of Coventry Cathedral, November 1940. The bombed-out shell of the 14th-century cathedral on 15 November 1940, the morning after the city was attacked.
© Science & Society Picture Library

But there was another factor limiting the effectiveness of Birmingham's response. The city's Chief Fire Officer, Alfred Tozer, had held the job since his predecessor, who was also his father, died in 1906. His deputy was his son. Tozers had been running fire services since the early 1800s, and the family's views had not changed a bit in that time. He would have nothing to do with the AFS. As for trade unions, Tozer told John Horner, "I'll tell you what my industrial relations are. If a man steps over the line in this brigade, I can get rid of him – like that," and he snapped his finger and thumb. "That's the only industrial relations I need."

Tozer's refusal to allow the regulars and the AFS to integrate meant that when Birmingham was hit, most of the city's fire crews were leaderless. He spent the night in his control room, muttering, "This thing can't go on – it must stop." And the next morning, in the small hours, Commander

Coventry's ruined streets on the morning after the raid.
© Rue des Archives / Mary Evans

Firebrace and Sir Arthur Dixon arrived from the Home Office in London and quickly arranged for the immediate retirement of Mr Tozer and the removal from office of his son.

Bristol – which had been considered relatively safe, and had taken in many evacuees – was raided on 24 November. The city's Art Gallery was destroyed and part of the University badly damaged. Once again, water failure hampered the efforts of the city's firefighters.

Cliff Latchem, an AFS pump driver, wrote an account of his experiences that night. He was one of a five-man crew that set out from Wells Road sub-station, heading for Bristol.

We coupled up the MP [major pump] unit to a Ford two tonner, forward control coal lorry with wooden sides, which was either loaned or hired to our local council; we had already put aboard our ladder and hoses . . . it was a good clear night, cold but no wind blowing . . . in front of us was a huge red glow in the sky from the fires in the centre of Bristol, and the bombs were still coming down.

We did not talk very much going in but to be honest I did not feel very brave under the circumstances, and also we were feeling very cold so on we went . . . down the Bath Road by Arnos Vale cemetery, where in the road were stones, shrubs, branches of trees and earth, which made our driving difficult. We were told later that a bomb had fallen into the cemetery and the stones were in fact gravestones. What with the glow in the sky and the fires, it looked like driving down into hell . . .

Then we approached Victoria Street and this was like Dante's Inferno. Raging fires in the buildings, debris all over the road, it was like driving through a tunnel of fires belching out of windows and doors from the tops to the bottoms of buildings . . . even the shop window glass was melting . . .

Having found their way to the Bridewell HQ in the centre of Bristol, the crew was detailed to drive back out of the city and deal with fires in the Temple Gate district; but

> *there was no water in the hydrants as most of the mains supply*
> *had been fractured by the bombing so we used the water in the*
> *[static] tanks and had pumped them dry by about 11pm.*
> *Although we were near to the River Avon, the tide was out so we*
> *could do no more . . . That night, there were two hundred people*
> *killed and six hundred and thirty-nine injured.*
>
> *This raid found a lot of faults in the organisation and one of*
> *the worst was having to go into the centre of Bristol to report; by*
> *the time of the next raid, this had been altered [and] you then had*
> *to report to Fire Officers on the outskirts of the city.*

Another large London contingent was quickly despatched
to Southampton when it was targeted during the last week of
November. On 8 December there was another massive attack
on London, and the LFB dealt with 2,000 fires. Ten men from
West Ham Fire Brigade died that night when a bomb hit their
station as they slept. Only auxiliary fireman Len Townrow was
awake when the bomb hit, on duty in the watchroom; he had
heard the scream of the bombers and was just about to wake
his comrades when it landed. He was the only survivor. He
took six months to recover from his injuries, and was then
judged unfit for further service and discharged from the AFS.

On 12 December it was the turn of Sheffield, another city
full of factories and workshops engaged in war-related
industries. The raid lasted nine hours. Although the fire
services could not save the city centre, they did succeed almost
totally in preventing fires from linking up with others.

Shortly before Christmas the Luftwaffe targeted
Manchester. The historian Eric Midwinter, whose father
Harold Midwinter joined the AFS in 1938, describes the attack.

> *There was a particular four days, ending on 23 December 1940,*
> *when there were four all-night attacks. As soon as it went dark,*

there was bombing right through till dawn, or more or less dawn, next day . . . ten-, twelve-hour bombing for four nights on the run, which is what's usually referred to now as the Manchester Blitz.

The Greater Manchester area [included] Manchester and Salford Docks, and Manchester Ship Canal, Bridgewater Ship Canal . . . and big munitions factories in Trafford Park, Metropolitan-Vickers, making Lancaster bombers, . . . so that was quite a target.

The city, according to the report in *Front Line*,

was caught off guard. Some of the shops and offices were well protected by roof watchers, who were able to cope with fire bombs, plentifully though they fell. Other buildings were quite empty, and the fire sentries elsewhere had to look on across the chasm of

Air raid damage, Sheffield, 13 December 1940. A bomb crater in one of the main streets. Photograph by F W Greaves. © Science & Society Picture Library

an intervening roadway, powerless and in fury, while two or three incendiaries slowly – so slowly! – burned their way through the roof and the top floor, until the whole building gradually roared into flame and was lost.

It was the fire brigade's first raid in their own city. The auxiliary firemen and their regular comrades had not yet had an opportunity to become welded into a single well-exercised unit, and the problems of mobilisation, command and water-relay were thereby the more formidable . . . Many blocks in the middle parts of the city were burned right out or severely damaged. All this marked no failure of heart or will on the part of the fire-fighters, but a lack of experience of what the most concentrated kind of fire raid could achieve, and of advance measures to cope with it.

It was on the last night of the raid, 23 December, that Eric Midwinter's home town of Sale was attacked.

That night, in Sale alone, there were ninety-eight fire calls. Apparently a stick of 200 incendiary bombs fell at one period, then a couple of hours later another 200 fell. The fire station was just at the back of the Town Hall, and Billy Clarkson, one of the firemen, he used to be up in the tower where they hung the hoses

---

Harold Midwinter (second row, seated, fourth from right) with Sale Squad.
© Eric Midwinter

to dry them, and they used to use it as a watch-out, and of course he was watching the fires in Trafford Park and reporting down the intercom – "Kellogg's is burning" or some other place – and suddenly he shouted, "Bloody hell, the Town Hall's on fire! I'm coming down." The Town Hall, Woolworth's, a couple of cinemas, a little factory, one of the churches, were very badly burned at that time.

My father was helping to fight the fire in Sale Town Hall and he and another fireman got out just before the big Town Hall clock fell right into the council chamber where the blaze was.

As a historian of popular culture Eric Midwinter was struck by a curious foreshadowing of this event in a comedy routine popular just before the war. In his monologue "The Fire Chief", the much-loved Robb Wilton (he of the "Home Guard" sketch that begins, "The day war broke out . . .") relates:

The last fire we went to, we, we hadn't much time to mess about . . . we'd a call through that the Town Hall was on fire. Well now, my foreman's a very decent . . . he's one of the best lads in the world, but no memory. Ooh, he's got a shocking memory. We dashed off, and when we were about half way there, we found we'd forgotten the fire engine. Well, of course, that meant turning back again . . .

The Mayor came rushing in, and he said, "Quick! Quick! For goodness sake – the left wing has gone and the right wing is going!"

I said, "Well, how's the centre-forward?"

I said, "What's the matter?"

He said, "Matter?" He said, "You stand there and calmly ask what's the matter, when any minute, the roof might fall in?"

I said, "The roof? Which roof?"

He said, "The Town Hall roof!"

I said, "Oh, Great Scott, you frightened the life out of me. I thought you meant this roof."

He said, "Quick, quick! Get your men together and do your duty!" Then he started to run out of the yard.

I said, "Wait a minute, wait a m–, as long as we're going we'll give you a lift."

He said, "No, no, no, they're expecting me back again."

Manchester firefighters complained they were offered a single cold sausage for their Christmas dinner. For those fighting the bombing in Southampton, Christmas dinner was hardly more festive. Harold Newell was sent there from north London to support his fellow firefighters. His daughter Lyn Nixon remembers:

It was Christmas Eve. And somebody appeared with some sausages, but they'd got nothing to cook them on. My father found a dustbin lid, which they scoured as best they could, and put it on a brazier and cooked these sausages. And he said it was one of the best Christmas lunches he could remember, because it was so lovely to have something hot.

Christmas itself was quiet, and a few resourceful fire stations were even able to organise a little festive entertainment. Merle Cartwright, then a child in south-east London, remembers how the AFS responded:

. . . when my father arranged a Christmas party in a local hall for children in our road. Money and goods were in short supply, but when Father Christmas appeared to give out presents all the girls except the very young were given a needlework box. These had all been made by men and women at the AFS station . . . My sister and I still have and use ours nearly seventy years later. It has been only in adulthood that we've appreciated the workmanship and long hours that must have gone into making so many, and I wish so much that they could have known in later years how valued their gifts to us still were.

Pat King, in Birmingham, recalls

fantastic Christmas parties held at the Fire Station, with party food – a real treat then – and presents. Lots of the presents were made by the firemen themselves and I clearly remember my father [an accountant by day, an AFS firefighter at night and weekends] making model battleships out of scrap wood and painting them "battleship grey".

But the Christmas break was the calm before the storm.

Christmas party at Bowes Park Fire Station, London, 1945. Harold Newell is third from right, back row, and his wife Kathleen is in the front row on the right.
© Lyn Nixon

Chapter 7

29 December 1940
The second great fire of London

In the City of London, near St Paul's Cathedral, key buildings were crammed together: offices, churches and warehouses (often filled with flammable materials), separated only by narrow passages and alleyways; and the Cathedral itself, which had huge symbolic importance and which Prime Minister Winston Churchill had decreed must be saved at all costs.

On the night of 29 December 1940 that square mile was directly targeted by the Luftwaffe, and the bombers left an area of destruction stretching from Islington to St Paul's. Around 1,500 fires were started. Many of the City churches designed by Christopher Wren after the Great Fire were destroyed or damaged that night. Other buildings erected after the Great Fire, such as livery company halls and part of the Guildhall, were also hit.

The City was an obvious target, yet, perhaps because of the Christmas lull, perhaps because German attention seemed to have been diverted to the provinces, it was not ready that night to face an attack. Offices and warehouses were padlocked, making the firefighters' task harder. It was Sunday and few firewatchers were about just after 6pm when 130 German aircraft suddenly appeared over the City and started to fling down incendiary and high explosive bombs. Fires were leaping from roof to roof, and explosives were smashing the water mains before the fire brigade could get there.

Firefighters laid a hose from the river to pump in water, and the cable was destroyed by bombs within half an hour. They laid more cable, but the Thames was at a very low ebb that night, and the firefighters had to manoeuvre their pumps down stone steps into the soft mud and walk through it, waist deep, to connect hoses to the fireboats. Some pumps became bogged down in the thick mud and had to be abandoned later when the tide came in.

By 8pm, 300 pumps were in the City, but their effectiveness was drastically reduced by the lack of water. More pumps were on their way; the decision had been taken to mobilise every available appliance, even at the risk of leaving other parts of London without cover. By 10pm, 1,700 pumps were working in and around the City, and only thirty pumps remained in fire stations to deal with anything that might happen elsewhere in or around London. But as their water supply dried up, firefighters were forced to retreat. Bulk water carriers were ordered in from outlying rural areas.

While the men manned the pumps, women were driving petrol carriers, canteen vans and staff cars into the thickest parts of the blaze, ensuring that the pumps had petrol to keep going. Driving vans laden with cans of petrol through the flames that night was about as dangerous a job as you could do. And, when necessary, they would step out of role to help control the huge, bucking hoses. "With that pressure, there was two feet of metal on the end that would take your head off," recalled one firewoman, Gladys Gunner.

When an area round Redcross Street had to be abandoned, pumps were left where they stood and firefighters made their way into an air raid shelter that led to an underground railway track of the Metropolitan Line, and followed the track until they came out at Smithfield. "There was fire in every direction," wrote one AFS man afterwards. "The city was turned into an enormous, loosely stacked furnace, belching black smoke. The sky above was red with the glow, and the smoke as it drifted upwards was reddened on its underside, but deep black when seen against the sky."

Harold Newell, an AFS firefighter based at Bowes Park Fire Station in Bounds Green, in north London, was with a crew

detailed to go to the City that night. As his daughter Lyn Nixon remembers,

> They suddenly realised that there had been bombs at both ends of this street and they couldn't get out. They were trapped inside this street. And the only way they could think of getting out was to go down in the sewers. And they went down in the sewers and managed to come up somewhere where they could get out. And he said that when he got home he just had to take his whole uniform off outside – the stench was absolutely incredible.

At some point in the night, Harold Newell found himself close to St Paul's. "He remembered standing there and thinking, 'If this goes down, we'll all go down.'"

His underground escape had a curious sequel. A few months later, Lyn Nixon says,

> he and the men who were on his fire engine with him were called in to a disciplinary action. And they said to them, 'You left a fire engine.' And they said, 'Yes, we had to, we couldn't get the fire

---

Harold Newell (back centre) and the crew at Bowes Park Fire Station in North London. © Lyn Nixon

engine out.' And the officer in charge opened up a German paper and there was their fire engine, with all this propaganda – London is finished, they're having to leave their fire engines, they can't get them out. Who had taken this picture and sent it to Germany nobody knows.

AFS man Hubert Idle from Wandsworth Fire Station was there too, and it was a night he never forgot. He dreamed about it sometimes, throughout the rest of his long life. It was dreadfully hot and bombs in the road were turning the tarmac into liquid. The police at first turned Hubert's crew back, saying it was too dangerous, but they persisted.

Two of Wren's churches, St Bride's and St Lawrence Jewry, were lost early that night, as their wooden roof beams collapsed, and six more followed later. Wren's greatest edifice, St Paul's itself, was one of the few buildings that had firewatchers, and vergers and choirboys were ready to throw buckets of water over the firebombs that landed in the roofs of the lower levels. But one incendiary landed high up on the dome, where no one could get at it – even if the firefighters had had enough water to use powerful jets, which they did not. Fortunately, the bomb slipped down the dome into the gallery that surrounds it, and buckets of water were quickly poured over it.

St Paul's was not saved yet. The streets all around it were on fire, whipped up and spread by a strong wind. A photograph taken from the roof of the *Daily Mail* building, showing the Cathedral shrouded in smoke, has become one of the most famous images of the Blitz. It was taken by Herbert Mason.

I remember [that night] only too well . . . In the distance through the smoke you could see the fires increasing, and as the evening wore on an artificial wind sprang through the heat caused by the fires, parted the clouds, the buildings in the foreground collapsed, and there revealed in all its majesty was St Paul's . . . Down below

in the street I went towards Ludgate Hill, which was carpeted in
hose pipes, a scampering rat here and there, a reeling bird in the
flames. The heat became intense as I approached St Paul's
Churchyard . . . embers were falling like rain and clattering on
your helmet. Cheapside was a mass of flames, leaping from one
side of the road to the other.

Back at my vantage point on top of the Daily Mail building,
where I was, I could see that this night I was going to obtain the
picture which would for ever record the Battle of Britain. After
waiting a few hours the smoke parted like the curtain of a theatre
and there before me was this wonderful vista . . .

Across the river in Lambeth, an onlooker saw

an unforgettable sight. The whole of London seemed involved, one
great circle of overwhelming disaster save in one corner . . . where

Firefighters in action © London Fire Brigade

the night sky was clear. One could not distinguish known buildings through the great clouds of smoke, except when there was a sudden spurt of yellow flames which lit a church tower . . . it seemed impossible that the City, that London, could be saved.

Suddenly and unexpectedly, at 9.45pm the "all clear" sounded. The City was still in flames, but it was a chance for firefighters to get to grips with it. The Thames tide turned, it became a little easier to get water from the river, and by 3am firefighters were starting to get the flames under control.

It was a dreadful night, but it could have been even worse. A hundred thousand incendiary bombs fell on London. Of 1,500 fires logged, all but 28 were in the square mile of the City.

Fourteen firefighters were killed. One of these was Frank Hurd, whom we last met fighting the blaze at Beckton Gasworks on the first day of the Blitz. Two others were Benjamin Chinnery and Herbert Blundell, AFS firefighters who died at Dingley Place, near the City Road. Chinnery was up on a ladder when a bomb blasted a wall, killing both men. Since Blundell's wife could not afford to bury him separately, Chinnery's wife agreed to them being buried together. A newspaper article headlined "Nation On The Cheap", preserved by the Chinnery family, reported that since Chinnery did not complete a full week's work, his widow did not receive a full week's pay.

Front Line the official Civil Defence account, described what happened:

A young fireman and his team, fresh from a successful struggle to keep the flames away from Dr Johnson's house [in Gough Square, off Fleet Street], lost their way in the smoke down a narrow street and happened upon a trailer pump crew working hard at a printing press building.

I thought when I saw them [reported the young fireman] that they were too near. Just at that moment a wall, which looked as if it was bulging dangerously, crashed down on them. As we looked round all we could see was a heap of debris with a hose leading towards it.

Seventeen awards were made for bravery, including four to auxiliary firewomen who had driven essential vehicles through a hail of bombs and a wall of fire.

St Paul's had not escaped unscathed: a bomb destroyed the High Altar, and there was damage to the crypt and to some of the stained-glass windows. Yet the great cathedral symbolised hope. In the words of the BBC reporter Robin Duff, "All around the flames were leaping up into the sky. There the cathedral stood, magnificently firm, untouched in the centre of all this destruction."

But while St Paul's still remained, eerily alone, almost everything round it had burned down – mainly because the cathedral had firewatchers and the other buildings did not. It was a testament to the need for firewatchers, and a Fire Precautions Order was rushed through Parliament requiring men between the ages of sixteen and sixty to register as part-time firewatchers and be on duty for 48 hours a month. This order was later applied to women as well.

Fifty years later John Horner wrote, "Of the many firemen who were killed that night, six died around St Paul's churchyard. In 1990 annual office rentals there fetch £550 per square yard. The widows of our dead were each granted £7 10s. to bury them."

Bomb damage, Cheapside, 1940 © London Fire Brigade

Chapter 8

30 December 1940–9 May 1941
Attacks on industrial cities and ports

By the end of 1940, raids in London had killed almost 13,000 civilians and seriously wounded 20,000 more.

At the start of 1941, it looked as though the raids were going to tail away. British defences were much improved by this time: ground-based radar guided night fighters to their targets, and the Bristol Beaufighter, with airborne radar, proved to be effective against night bombers. London was far better defended, with anti-aircraft guns and searchlights, increasingly radar-controlled, improving accuracy. From the beginning of 1941 the Luftwaffe's monthly losses rose, from 28 in January to 124 in May.

London was bombed on four nights in January, and Portsmouth and Plymouth were hit hard; pumps and crews were sent there from London. Even so, as a firefighter reported afterwards, "The constant cry was for water, water, and still more water . . . It was eventually relayed from a distance of over three miles, measuring the roundabout route made necessary by bomb damage."

In the early months of 1941 the Luftwaffe concentrated on industrial and port cities, following a directive issued by Hitler on 6 February. Targets included Coventry, Southampton, Birmingham, Liverpool, Clydebank, Bristol, Swindon, Plymouth, Cardiff, Manchester, Sheffield, Swansea, Portsmouth, Avonmouth, Barrow-in-Furness, Belfast, Hull, Sunderland and Newcastle.

The attack on Cardiff on 2 January began with a heavy fall of incendiaries, but police and fire guards were quick to extinguish them, and there were no really large fires. Llandaff Cathedral, however, was the victim of a landmine that fell

close to it, blowing the roof off the nave, south aisle and chapter house and also damaging the spire and the organ. Of all British cathedrals only Coventry was worse hit.

It was so cold during the raid on Bristol on 3–4 January that not only did the fire hoses freeze: the firefighters' tea, provided by the WVS, froze in the cups. "Two houses might be seen side by side, one in flames with the firemen at work on it, the other hung with long icicles where the streams of water had splashed and frozen."

Swansea was the subject of an onslaught from the air for three nights, 19–21 February, when 800 high explosive bombs and 30,000 incendiaries were dropped. Although the city was a significant target for the enemy because of its docks, much of the central business and residential area was destroyed.

Belfast was poorly defended, with just seven anti-aircraft guns, and when the city was attacked on the night of Easter

Between 19 and 21 February 1941, Swansea was attacked, 230 citizens were killed and most of the city centre was destroyed. Picture courtesy of the South Wales Evening Post, Swansea

Tuesday 1941 it suffered the greatest loss of life in a single raid on the United Kingdom outside London. The city's guns stayed silent for fear of hitting defending RAF fighters – which had never actually been scrambled. The 1942 Civil Defence report *Front Line* described the scene:

> *When morning broke, parts of the city were a gruesome sight. The ruin-fringed roads were blocked by heaps of smoking debris and acrid-smelling craters. Water ran through the rubble, gas mains spouted fountains of flame, and where the fire-fighters were still at work every now and again a wall crashed.*

Both then and in raids the following month, the city's fire service was supported by reinforcements from mainland Britain and the Republic of Ireland.

Glasgow and Clydebank were hit on 13 and 14 March. At the start of the war, Clydebank had been considered likely to be bombed, and the majority of the town's women and children were evacuated; but when the raids failed to happen, many evacuees returned to their homes.

In two nights of fierce bombardment, Clydebank was devastated by 1,650 incendiaries and 272 tonnes of high explosive bombs. It was reported that only seven of the town's 12,000 houses escaped damage. Five hundred and twenty-eight people were killed and hundreds more were seriously injured. Firefighters worked for 50 hours continuously to put out fires in a group of oil tanks, according to *Front Line*,

> *. . . and at the end there were ninety-six high explosive bomb craters in the limited area over which they had fought. They waded through the moats round burning or threatened tanks; they climbed up the ladders and blacked out jets of burning oil gas pouring from holes in the crown of the tanks; they worked near the sides of the tanks under the blazing drips falling from above. They hosed one another as they worked to make it possible*

to go on. Not only did they save a good number of the threatened tanks but they extinguished some that had caught fire and been burning for as long as two days – an excellent rare feat.

Rushing men and equipment between London and provincial cities caused all sorts of problems, not least of command: London-based commanders tended to want to run things their way, and their orders were sometimes counter-manded by local commanders. But the relative lull in London did allow time for the London Fire Service to consolidate, and to release men from front-line duties so that they could attend courses on such matters as breathing apparatus and turntable-ladder operation. It also allowed the LFS to bring into service some new equipment: purpose-built vehicles for towing the trailer pumps, a job which until then had mostly been done by London taxis, and steel piping for emergency water supplies, which could be laid at the rate of a mile an hour. The LFS was also provided with large numbers of portable canvas dams, and static water tanks were installed in the basements of many bombed-out buildings.

London was the target for further flurries of raids in March and April. It was not the nightly bombardment it had known in the Blitz, but there was a particularly damaging raid on 16 April, when 685 German planes dropped 890 tons of high explosive and 4,200 canisters of incendiaries on central and southern districts of London, killing over 1,000 civilians and wounding over 2,000, and damaging eighteen hospitals and thirteen churches. The LFS managed to bring the fires under control by morning without calling on provincial fire services for help.

St Paul's was hit again, and saved by firewatchers again. An AFS sub-station in Chelsea was hit, three firefighters killed and others injured. In Victoria, a turntable ladder was rushed in and erected to aim water at the roof of Christ Church, but before it could be used it was hit by bombs. One firefighter was killed at once, two died soon afterwards, and five more were

seriously hurt, but the man right at the top of the ladder was only knocked unconscious. The ladder was blown off the turntable but remained intact and lodged on the wall of the church. When he regained consciousness, the firefighter climbed down the ladder. Three days later, the enemy returned and 712 planes dropped 1,026 tons of high explosive and 4,252 incendiaries.

In March and April, Plymouth was targeted on seven nights. The Guildhall, Law Courts and other public buildings were destroyed, as were the shopping centres of Plymouth and Devonport, and thousands of houses. "Most of this damage," according to *Front Line*, "was done by overwhelming fires which presented problems of water provision and reinforcement that were almost insoluble in the conditions then and there obtaining . . . It would be a different story to-day." The city's own fire service was unequal to the problem,

Plymouth. Mr Widdicombe walks through the bombed streets to salvage what he can of his belongings, 1940 © Robert Hunt Library / Mary Evans

and reinforcements were summoned from eighteen outlying towns; pumps were also obtained from naval and military establishments. On the last night of bombing, the *Front Line* account asserts, "there were no fewer than 12,000 ordinary citizens organised and brigaded as fire-bomb fighters, with who knows how many unorganised and uncounted to add to their number."

In early May, it was Merseyside's turn for seven nights of unrelenting aerial assault.

> *The policemen and firemen who guarded the [Merseyside] docks were perhaps under as fierce an attack as any men in the whole course of the onslaught on Britain. High explosive and incendiaries fell in great weight upon the dock basins, the quays, the ships moored at their sides, and the store sheds hard by . . . Firemen fought all night to check the fires on a blazing munitions ship . . . the threat from flames and falling embers, great as it was, being less than the chance of immediate explosion at their sides if the flames moved too fast for them.*

Most of Birmingham's fire service was sent to Liverpool, and the LFS rushed appliances up to cover in Birmingham. Harold Midwinter went to Liverpool with the Sale fire brigade, and remembered having to deal with a fire in the Tate & Lyle factory: "Syrup was streaming everywhere, on fire, and we were having to wade through hot syrup and blistering sugar."

Hull was easily reached and readily identifiable by German bombers coming from the direction of Denmark, and it endured almost 50 raids between summer 1940 and May 1941, as well as others later. The city, said an official account the following year, "is scarred from side to side and end to end."

Fred Bulmer was an AFS, and later NFS, firefighter in Hull throughout the war. "The only mention he made of his wartime service," writes his son Eric,

was when we saw one of his ex-servicemen (an insurance premium collector) pedalling past the house. He told me that one "bad" night he and this chap had set up a hose, presumably on a stand of some sort, and had taken shelter from the bombs. The order came through to them to move the set-up. The other fireman said, "Would you do it, Fred? I'm only timid." And so my dad did it.

I must have spent quite a few nights in the air raid shelter at the bottom of the garden but don't remember as I was not born until January 1940. I do remember Dad playing the "take shelter" and "all clear" on his violin, on request.

*

By May 1941 around 300,000 houses had been destroyed and over 20,000 civilians killed in the London area. More than a million houses had been damaged, and 375,000 people had been made homeless.

Lord Steet in Liverpool after the heaviest raid the city suffered during the war. A fire appliance sent from Blackpool is on the right of the picture.
© Science & Society Picture Library

Stephen Spender talked to dozens of wartime firefighters and built up this picture of what happened when the call came.

Everything is prepared, so that the crew can jump into their vehicles and go straight where they are needed. The firemen are either dressed in their leggings, and wearing their belts with spanner and axe, or their gear is already on the appliances. When the bells go down, they dash out of the station into the yard, jump into the appliances, and fumble their gear on in the dark. The machines race through the streets with their bells clanging. When the fire engines arrive at the scene of a fire the first task that confronts the firemen is to find water. This is by no means easy in the dark. It was especially difficult in the early days of the Blitz when the hydrants were not clearly marked. Having found their hydrant . . . the firemen must fix the hydrant head to it and then the coupling of the end of the hose . . .

Two firemen may have to stand for hours holding the "branch" (or nozzle) from which the water proceeds, pulling them backward in the opposite direction to the stream of water . . .

The main feeling of the Blitz was exhaustion . . . For most of the firemen and rescue men it meant working, two days out of three, from blackout til six or seven in the morning. On the average there was about 11 hours work a night. The men were drenched through doing this work, tottering with sleep, extremely cold, and bruised with the weight of the material they had to carry.

John Horner wrote:

What has not been told in the press is the complete lack of preparedness which left men isolated for hours without food or drink, which condemned men who had been wet through for days to return to their stations and turnout again, still in wet clothes.

With the whole of the London Fire Service and the entire London Region completely mobilised, there were fewer than six mobile canteen vans in commission.

80

Liverpool, May 1941. Firefighters work in the rubble of the ruined church, directing a jet on still-smouldering debris. © Science & Society Picture Library

Chapter 9

10–11 May 1941
The longest night

The last major air raid on London, on the night of Saturday, 10 May 1941, was also the most destructive. Two elements dreaded by firefighters – a full moon and a very low Thames ebb tide – were to coincide and make firefighting even more hazardous. German bombers would use the bright moonlight to follow the Thames and spot their targets; the low tide would mean that firefighters could not rely on Thames water to feed their hoses. It was to turn into London's longest night of bombing.

It was two weeks since the Luftwaffe had mounted its last extensive 1,000-ton bombing raid against London, and the city's inhabitants had begun to believe that the Germans had switched their targets elsewhere. Fire services had been sent from the capital to help those now facing the Blitz in Birmingham and the Midlands. But Londoners were cruelly disappointed, as the German bombers, in retaliation to the British bombing of Bremen and Hamburg, launched what would be one of their most devastating raids against the city. Some of London's most important buildings would be damaged, including the Houses of Parliament, St James's Palace and the British Museum. The Queen's Hall, one of the city's major concert venues, would be destroyed, as would much of the area around St Paul's Cathedral.

Saturday had been cold and clear. It was Cup Final Day, with Arsenal playing Preston North End at Wembley, and 60,000 supporters had taken a few hours off from the war to support their teams. But as the fans cheered them on, the Chief Officer of the London Fire Service, Major Jackson, based at LFS headquarters in Lambeth, received a call from the RAF

monitoring unit to warn that the radio beams used by Luftwaffe bombers to guide them towards their targets were set to intersect at West Ham. Jackson told his men to stand by.

As people made their way home, enjoying the light of the two extra hours of Summer Time that had been introduced the previous week, the fire brigades prepared for the worst. Two thousand five hundred pumps were to be made available from

A bomb site next to St Paul's Cathedral, London, seen from Bread Street in the 1950s
© Mary Evans Picture Library

London and its county boroughs such as Croydon. The 50 LFS appliances sent up to Birmingham were placed on standby to return if, as expected, the air raid was a big one.

As night approached there was no sign of enemy aircraft, and no warning "red alert" was sounded. Routine dusk patrols carried out before the sun set at 9.36pm had found no trace of anything unusual. In the West End, audiences left the cinemas at 10pm, taking advantage of a new closing time that gave them twenty minutes to get home before the blackout. Several thousand firefighters waited at their stations across London. All seemed quiet. It was a cold night. The silver moonlight shone down on the great expanse of mud revealed as the Thames ebb tide swept out. By 10.30pm the river had dropped eighteen feet below the high-water-level mark.

It was the duty of each London borough to sound the alarms once alerted of any raid by Scotland Yard. The first sirens of the night sounded at 10.45pm as the first twenty Luftwaffe Heinkel "fire raisers" flew over the city, dropping their incendiaries. The first string clattered down on to the east casements of the Tower of London, setting light to the Constable Tower. Hundreds more pelted down, lighting up the city.

British fighter planes, which seem to have been caught off guard, failed to appear, as the Heinkels, unchallenged, dropped their heavy explosives indiscriminately. They were followed within the next twenty minutes by wave after wave of German aircraft, guided by the bright moonlight, the incendiaries and later the glare from the fires stretching across London from Dagenham to Putney. About 515 German planes, some of which would return at least three times to their bases in France, Belgium and Holland to reload and return, mounted one of the most destructive raids of the war.

West Ham with its docks had quickly been confirmed as a target, as it was one of the first to be bombed. Incendiaries followed by bombs rained down on it, and on the previously bombed Silvertown and the Surrey Docks. The railway

marshalling yard complex at Stratford was also among the first areas to be hit. Local volunteers at West Ham, used to the raids, dashed to put out the fires, telling the regular firefighters to "go ahead and deal with the bigger fires". But this time the conflagration was too much for them. All the available pumps in the area were switched to West Ham.

Arriving to inspect the damage, Major Jackson, realising the impending threat to other areas, deployed some of the pumps towards Whitechapel. Each of his senior officers was sent to take control and co-ordinate the pumps as they arrived at the various bomb sites. The LFS appliances that had been sent up to Birmingham were recalled as it became evident that the bombers were spreading their loads across London.

Fresh fires broke out quickly and spread along the banks of the Thames on a ten-mile front from Barking to the edge of the City, as more and more German bombers, following behind the first Heinkels, dropped their bombs.

Many of the strategic water mains were damaged and out of action, including the principal City trunk main connecting the Thames near Cannon Street railway station to City Road, and the West End main connecting the Grand Union Canal at Regent's Park to Shaftesbury Avenue. The flow from street water mains was depleted by the fracturing of the trunk mains.

Fire crew at Edmonton Fire Station, 1940. Margaret Hughes writes: 'My father, Bert Wornham (centre group second from left), was an engineer with the National Fire Service, not a firefighter, but he was living and working in the East End all through the war, making sure the vehicles and pumps were working.' © Margaret Hughes

The mains had been expected to generate some 30,000 gallons a minute, but jets of water directed by firefighters at blazing fires faltered and died.

The lack of water reached crisis point as water from the Thames, which could also be used to back up the systems, as well as being used to fight fires along its banks, also began to dry up. The firefighters had always dreaded the effect that a low tide would have on their chances for containing fires, and they had been right. Water directed from the fireboats, lying 50 yards from the Thames banks, slackened into thin streams. Desperate for water, crews were directed by their water officers to public swimming pools, canals and ponds. Throughout the night firefighters tried to get every possible drop of water, some setting up their appliances in flooded bomb craters or sewers.

Not all of the bombs were designed to explode at once. Some were specifically aimed at firefighters. These were delayed-action parachute mines that fell on the men as they rushed to deal with the fires. Thirty-eight of these had earlier fallen on West Ham, and now more fell over a wider area. Abraham Lewis, an AFS man stationed in London's East End, was one of the men killed by an incendiary. Called to deal with the fire at Trinity House in Tower Hill, he was trying to connect his hose to a hydrant when he was hit by an incendiary bomb that dropped on his back, even though it did not ignite – it shattered his spine.

By 11pm sirens were sounding in Westminster, and two minutes later in Kennington as the raid moved over central London. It was now evident that this raid was going to be one of the worst that London had faced.

Used to the Blitz, and having had a respite of several weeks, Londoners had thought that the first break in bombing, earlier that night, meant that it had stopped. Now they heard the drone of planes and looked up at the skies to see over a hundred German bombers, followed by a hundred more half an hour later. As the fires spread towards them, they made for their air raid shelters or, gathering belongings, ran towards areas not yet affected. The sky was now painted red. Over the Thames, flames reflected on the undersides of the silver barrage balloons flying high above the river.

Ernie Baron, then eleven years old and living in Lambeth, described the scene: "It was all nothing but red, everything was, fortunately our road wasn't hit but we could see the fires light up, there was also this terrible sound of droning."

Chief Superintendent C. P. McDuell was the Officer in Charge of the LFB's A Division, which covered the West End. He was waiting in the watchroom at Manchester Square for the first major incident. Told that the Houses of Parliament had been hit, he dashed to the site to discover that the roof of Westminster Hall and its adjoining buildings were on fire. A large unexploded bomb was found in the basement of the Victoria Tower and a fire had broken out in the Commons Chamber. Flames were beginning to leap up from St Margaret's Church opposite. The battle to contain the fires in these historic buildings was given precedence, because of their propaganda use. Other historic buildings would not get the same service, as the odds stacked up against the firefighters.

The night was rapidly descending into one of terror for all Londoners. The city was lit up with fires, which made targets like the Thameside docks and warehouses, and many of London's most famous buildings, easier for the bombers to spot. Many of the main road routes had been severed, causing major obstacles for the fire services. Several bus stations had also been hit; the transport system was going to be one of the major victims of the bombing. Southwark Bridge, a potential target, proved hard to hit; a cluster of incendiaries missed the bridge and landed on Southwark Fire Station, putting it out of action.

In just eighteen minutes after midnight, AFS women operators in the main central control room logged no fewer

Fire at Waterden Road, Homerton, 1940 © London Fire Brigade

than six potential conflagrations. Fires were burning virtually unchecked in Battersea, Moorgate and the City Road. Queen Victoria Street, which ran from Cannongate down to Blackfriars, was in flames. The Temple Church and Serjeant's Inn were ablaze, likewise several Wren churches, among them St Andrew, Holborn, and St James Garlickhythe, as well as Nicholas Hawksmoor's St George in the East, in Cannon Street. The British Museum was hit, as was the Great Synagogue and the Queen's Hall, which that afternoon had hosted a concert conducted by Sir Malcolm Sargent.

Many of London's main line stations, including St Pancras and Cannon Street, had been bombed. A huge fire was burning in the vaults below Waterloo Station's platforms as the paving stones buckled and melted, the asphalt like sticky gum. Hundreds of Londoners had taken shelter in the station's gentlemen's lavatory, where they remained safe, protected by the marble walls, whilst bombs fell not only on the station, but on The Cut, the street running alongside it, and on the bonded warehouses in York Road. Gallons of alcohol stored in the railway arches' warehouses poured on to the floor, and by 1am the arches were aflame.

Harold Newell from Bowes Park Fire Station remembered the burning bonded warehouses. "All the firemen were there, fighting the fire. Then, before you could turn round, the Customs and Excise men were all over the place." They were there, not to help fight the fire, but to check that none of the warehouses' contents escaped the fire in the pocket of a firefighter's uniform.

"Mac" Young from Paddington Fire Station was one of the firefighters drafted in to fight the Waterloo Station fire. He remembered arriving at the station, parking his pump in York Road and hearing the threatening sound of the bombs falling on them: "When we got to the entrance to Waterloo the fire under the station was terrible. A string of bombs came down, I seem to remember three. One, two, three and if the noise got

louder with each one you knew the next was coming your way."

The impact of high explosives would be described by firefighters as like that of a "troll hammering against the surface with a giant sledgehammer". The firefighters from Paddington, helped by ten pumps from outside London, spent most of the night fighting the fires in the vaults. Provincial crews brought in from outer London arrived within a few hours of the bombing starting; unused to the Blitz, they had to be directed by the more experienced London hands on what to do. London firefighters laughed as the provincial crews arrived in their spick and span uniforms, but the bravery shown by the newcomers, once they had adapted to what they faced, won them a new respect.

Firefighters from 101 Waterloo Fire Station fought to contain the fires in The Cut as bomb after bomb landed on the

AFS firefighters in action at Queen Victoria Street, London, 1941
© London Fire Brigade

shops. Told there was no more water available, Fred Cockett remembered where some pumps might be stored and took action: "We knew we had to stop the fire spreading down The Cut. I got the staff car to central London and climbed into the first hose-laying lorry, with a key still in its ignition, started it up and drove back to Waterloo." The hoses were connected to a pump brought in from Kent, and water from the Thames was directed into a quickly constructed canvas dam.

The fires in Clerkenwell and in the City were spreading. By 12.25am all telephone links with the City, Holborn, Shoreditch and Bethnal Green had been severed. A warehouse next to the London Fire Service District headquarters in Roseberry Avenue, Clerkenwell, received a direct hit, and seventy-five sub-stations were cut off from their control. Valuable time was lost before contact could be re-established. Communications became more difficult as the night wore on. Links over the Thames bridges from London Bridge to Lambeth were cut as they were badly damaged. Fallen buildings blocked roads and 700 gas mains were fractured, adding to the fires.

Few areas were spared as bombs poured down on London. In the north of the city, Purcell Street in Islington was hit; in the south, Cunard Street, Southwark; in the east, Redmead Lane, Wapping; in the west, Notting Hill Gate. Putney Hospital in south-west London was also hit. Bermondsey was bombed for a second time, as was the City.

Two fire guards were killed in Buckingham Gate as they fought a blaze at the Duchy of Cornwall offices. Fifteen people were killed in Greenwich, and in Ebury Street, Victoria, Frank Gough, a 64-year-old fire guard, was killed when the steeple of St Michael's Church crashed down on him.

Henry Stedman, an AFS man sent to a fire at a warehouse near Cannon Street, witnessed a terrible scene.

No words can describe the inferno that raged in London that night. Overhead was the ceaseless drone of the bombers, the air was rent by thud after thud of heavy bombs, mixed up with the rattle of the incendiaries. Close to the railway station we saw a rather unnerving sight, a huge bomb crater that stretched across the road, and lying in the bottom, twisted and buckled almost beyond recognition, a tender with its pump. Without hope we speculated on the fate of the crew.

Throughout the night, the City of London burned, as exhausted firefighters did their best to contain the fires. Incendiaries dropped by Cannon Street Station around midnight had spread quickly, starting a conflagration that threatened St Paul's Cathedral to its north and the whole of Queen Victoria Street. Mansion House had lost all its windows as firefighters battled with fires, dousing incendiaries with their stirrup pumps. The art deco Chamber of Commerce had already burnt down, as had St Nicholas Cole Abbey and St Mary-le-Bow. The Salvation Army Headquarters was now also threatened.

Major Jackson ordered a fleet of lorries to bring water down to Queen Victoria Street from the Regent's Canal. Each lorry slowly worked its way down through the burning narrow streets to deliver 1,000 gallons each into a 5,000-gallon steel dam. Until the Thames rose again it would be the only way of saving the world's biggest telegraph exchange, Faraday House. The north-east block of Faraday House, known as the Citadel, was also Winston Churchill's emergency bunker, and was of vital importance to the future planning of the war. Two cabinet ministers were in residence in the bunker that night – Sir John Anderson, after whom the shelters had been named, and Ernest Bevin, Minister for Labour. It was essential to save the exchange if Britain was to remain in contact with the rest of the world. The LFS was less worried about the two Ministers.

Major Jackson looked on as his men fought the flames. If the relay failed to work, he had received orders to dynamite

the street, causing a natural firebreak. It was to be an action of last resort, as no one could be sure that the explosion would not damage the foundations of St Paul's, just 100 yards north, but it would save Faraday House.

The battle for Faraday House would remain in the balance until 9.30am, when additional firefighters, whose leave had been cancelled hours earlier, augmented by firefighters from

Attacking a fire with a stirrup pump, late 1940
© Robert Hunt Library / Mary Evans

outside London, began to win control. Faraday House and St Paul's would not be declared safe until 6pm.

Some of London's oldest areas around Fleet Street and Holborn were destroyed as flames licked their way down the narrow streets, burning buildings that in some cases had survived the Great Fire of London. The Old Bailey had been hit, as had Gray's Inn Hall and five livery halls. St Clement Danes, the RAF church at the top of Fleet Street, was set alight. So too were many of the streets running down to the Thames.

The FBU's new offices in Chancery Lane of which John Horner was so proud – he had got them cheap because so many firms were moving out of London – were completely destroyed, and he was lucky not to have been killed: he had changed his mind at the last moment about sleeping in the office that night. From that time on, the bombs seemed to follow the FBU around – they were subsequently bombed out of offices in Hampstead, then Holborn; and a bomb damaged, but failed to destroy, the union's final wartime home in Bedford Row.

As water supplies dried up, firefighters abandoned the attempt to put out burning buildings and concentrated instead on stopping the spread of the fires. Throughout the night they fought to save hospitals, buildings vital for war work, and those where there was a risk of heavy casualties. Firefighters on the roof of St Thomas's Hospital turned down the pleas of the verger from Lambeth Palace to put out the fire on its roof. The Archbishop, who was in residence at the time, was left to his own devices. Eighty people had been brought into the hospital's casualty department that night. Masses of debris had fallen down the lift shafts into the basement, where fifteen of the most badly injured had been put in a hastily created temporary ward. A large pipe had also burst, causing flooding and a loss of much-needed water. Firefighters worked frantically to pump it out.

Further south of the river at the Elephant and Castle, the

night had started more quietly. The area had been badly damaged in earlier raids and although incendiaries and bombs had fallen nearer to the river, south Londoners hoped they would not be targeted again. But as midnight approached, the bombers plastered their bombs over the south London boroughs. People living in the area who had gone outside to watch now dashed towards their shelters, as did those living in Camberwell, who ran towards the Elephant and Castle, where they were confronted by the flames beginning to lick around the buildings.

Men at the Westmoreland Road operational station scrambled into their fire gear as clusters of fire bombs dropped on their area. A stick of high explosives fell in to R. White's mineral-water factory in Albany Road, sending debris into the fire station's yard. Chemicals from the factory escaped into the air, adding to the nauseous gases already being inhaled by the firefighters. Blocks of flats were brought down as streets in the vicinity went up in flame.

A string of fires had joined up, forming a ring of flame. Streets bounded by Walworth Road, New Kent Road, Newington Butts and The Causeway were a mass of fire. Every hydrant was emptying or already empty. The first pumps arriving on the scene were set into a 5,000-gallon emergency dam near Spurgeon's Tabernacle, whilst water was brought in relays from the Manor Place Baths. The Tabernacle itself was already on fire and much of the water from the dam went to try and put that out.

Superintendent George Adams, head of the Elephant and Castle District, sent three pumps up the road to the bombed-out Surrey Music Hall at St George's Circus, where previously, to much criticism from the London County Council authorities, Divisional Officer G. V. Blackstone had installed an emergency supply of over 200,000 gallons of water in its converted basement. There were now 500 firefighters at the Elephant and Castle fire. Sixteen were sent up to run the pumps at the Surrey Music Hall. At 2.15am a stick of bombs blew apart five three-storey buildings in London Road, and seven minutes later a high explosive exploded among the firefighters at the Music Hall. All of them were killed. Access to the water dam was cut off. The fires, like those on the other side of the river, began burning uncontrollably.

Water brought by relays from the Thames provided the only way of fighting the fires. But collapsing buildings and red-hot stonework and timbers had fallen across the lines of hose that lay along the routes of the relays to the worst fires at the Elephant and Castle, St Paul's, Whitechapel and Westminster. Firefighters patrolled the lines trying to keep the water flowing. Some relays were stopped and had to be relaid because of the obstructions. The walls of a blazing building in Newington Butts collapsed, burying newly laid hoses. By 5am, nine miles of hose from the Thames at London, Waterloo and Westminster Bridges were delivering water to the Elephant and Castle fire zone. Hoses had been laid around the burned-out skeletons of three fire engines destroyed outside the Surrey Music Hall.

Fireboats out in the Thames plied up and down the river all night. Sometimes they had had to use their own powerful water jets. The small vessels had sought shelter under the bridges or just off the mud, feeding four tons of water per minute on to the shore and towards the fires. But there were far more pumps seeking water than the fireboats could hope to supply.

Dewsbury Dessau, an AFS man who was working as a crew member on a fireboat just below St Paul's, described the sound of the fire raging above him:

Way up above there was a muffled crackling like fists beating on a sheet of zinc, and a crescendo whine and a fresh fire flared up on the shore. The wind seemed torn the other way, and up from the very bed of the river came a trembling roar and boats shuddered, buckled and shuddered again. On the downstream side of London Bridge another hissing whine brought back an even louder echo, a rushing roar culminating in a tearing, slithering sound of rubble falling from a height.

As dawn broke, the Thames waters were again rising and more water flowed through the relays. Although the fires at St Paul's and the Elephant and Castle, and a hundred other places, were still not under control, the "all clear" finally sounded just before 6am.

Firefighters were almost asleep on their feet; some, drained of energy, were unable to stay awake and lay down to sleep in the streets. Few had taken any refreshments. Henry Stedman was later to write: "Of all my many fires I think that one was the worst. My feet ached so badly I could hardly stand, and I felt so tired and exhausted that when we were at last given a meal it required concentrated effort to eat it." At 6am Jackson cancelled all officers' leave and they, together with firefighters brought in from the provinces, took the places of exhausted and injured firefighters. Gradually they were able to subdue some of the smaller fires. The RAF, which had been mobilised for some hours, had been picking off the German fighters and bombers. No longer could the Germans rule the skies. With the break of dawn and the skies lit up, German planes became targets for the defences.

Dewbury Dessau on his fireboat recalled the effect of daylight: "Nosing on the making tide towards Blackfriars, I was suddenly conscious of the brittle lightness of things. A red blotch passed over the bridge above. It was a bus and it meant morning."

By mid-afternoon Jackson's officers were able to report that the City and the Elephant and Castle conflagrations were coming under control and that St Paul's Cathedral was safe. The fire at Waterloo Station burned for another four days before the LFS was able to bring it under control. Many of the

---

Firefighters at Queen Victoria Street, London © London Fire Brigade

fires carried on burning for days. The last LFS pump was not withdrawn until 22 May.

More than 500 aircraft bombed London that night. Seven hundred and eleven tons of high explosives came down, including parachute mines, and 86 tons of incendiaries. Although this was less than the previous April bombing, the fact that there was both a full moon and a low tide had helped the Luftwaffe to cause the greatest damage of the Blitz.

Seven hundred acres of London were damaged by fire. Thirty-five factories producing war equipment were bombed. Two thousand two hundred fires were recorded, including nine conflagrations (fires burning out of control); twenty major fires needing over 30 pumps; 37 serious fires needing up to 30 pumps; and 210 medium fires needing up to ten pumps. Some 1,364 people were killed and another 1,616 were seriously injured; 12,000 were made homeless.

For the defence services, this night emphasised with terrible clarity two problems that they were not, at that point, fully equipped to solve, as *Front Line* explained

> *One was how to fight fires when main water supplies failed – the problem of emergency water. This could be solved only by an elaborate constructional programme which did not approach completion until much later. The other was how to concentrate defensive forces on the ground at a speed to match the intensity of concentration which the enemy could sometimes secure for his attack from the air. This was the problem of mobilisation and reinforcement. To solve it required radical reorganisation.*

And that solution would come – in just a few months' time. It meant thinking the unthinkable, and ministers had already thought it. It was not yet public knowledge, but two days before the raid, the cabinet had agreed in principle to nationalise the fire service.

Chapter 10
12 May 1941–1 February 1942
The fire service is nationalised

On 11th May, as the firefighters put out the night's fires, as the wardens and rescue parties dug for the buried, as the ambulances rushed the wounded and dying to hospitals and the mortuary vans collected the dead, as people once again swept the glass from their door-steps and the dust and rubble from their floors, no Londoner could know that it was over.

But it was, or almost, for the time being at any rate. There was one more major air raid on a British city on 16 May, when 111 bombers attacked Birmingham. Thereafter, England would not experience a significant air attack for about a year and a half. Hitler had a far more prized target. In the following month, Operation Barbarossa was launched – the attack on Russia. The huge military force needed for this attack included many bombers, and two thirds of the German military was to be tied up on the Eastern Front for the duration of the war. The imminent threat of an invasion of Britain had passed.

By the end of May 1941, over 43,000 civilians, half of them in London, had been killed by bombing and more than a million houses had been destroyed or damaged in London alone.

It was clear that Britain's fire services badly needed co-ordinating. Their response had been made less effective by the divided command. Sir Arthur Dixon, head of the Fire Services Division at the Home Office, said, "Nationalisation is impossible. The whole of history is against it."

But the firefighters, and their union, wanted it, and it was clear to them, if not to Sir Arthur, that without nationalisation it was impossible to mobilise fire appliances on a large scale at great speed. "If heroism and devotion to duty and self-sacrifice in themselves could constitute an efficient fire service, there would be no need of any reorganisation," John Horner told the 1941 FBU conference. But "we are all serving firemen, and we know the mistakes that have been made, and we know how we, as firemen, have had to suffer under the inefficiency, the maladministration and the blunders which have been made by local authorities." There were, he reminded his members, 1,400 fire authorities, "and when the government has produced a memorandum about the fire service, they have usually put 1,400 different interpretations on it." At this point, according to the minutes, delegates called out "hear hear".

On 18 April 1941, Home Secretary Herbert Morrison called in Dixon and Firebrace for a four-hour meeting that ended at 2am. They sketched out the idea of a national fire service.

It was a huge task. Many of the 1,400 fire authorities had both regular and AFS brigades. They all had to be put under overall control. The country was divided into 33 (later increased to 43) fire areas, grouped in eleven regions. Mobile divisions were created, with their own transport, field telephones and equipment. The government took on the cost instead of each local authority paying for its own service. The regional regular fire brigades and the AFS were merged.

On 13 May, in a city still smoking from the heaviest raid London had ever experienced, Morrison introduced the Fire Service (Emergency Provisions) Bill, which received the Royal Assent just eleven days later; its effect was to nationalise the fire service.

The National Fire Service (NFS) was officially created on 18 August 1941. Morrison issued the first orders of the day:

> . . . You stand in the front rank of our defence against the menace of air attack. You have faced tasks such as no fire brigades in the world's history have ever been called upon to perform . . . The present change is being made in order to weld the many local fire services into a single national service, which can be more

effectively organised, trained and directed for large-scale firefighting operations.

He told them to "train, organise, practice and be ready."

The benefits were felt at once, wrote Stephen Spender. The most amateurish of the officers – those who were in charge because of their social status in the community – were removed and replaced, and a lot of the spit and polish was abolished. No longer might the chain of command involve the town clerk, the Chairman of the Main Drainage Committee, or the Master of Corpus Christi College, whether or not he happened to be awake.

"The setting up of the N.F.S.," wrote W. Eric Jackson,

the appointments of the new commanders and other senior officers, the settlement of the areas, the allocation of personnel, equipment and appliances, the structure of chains of command, and the drafting of the requisite statutory regulations and directions were carried out in less than three months. Besides which, premises had to be found, new uniform badges and marks of rank devised and issued, and standard pay, hours and conditions of service had to be arranged. The change-over was described in Parliament as one of the finest pieces of large-scale organisation carried out in this country.

Aylmer Firebrace, though he had predictably opposed the idea, made good use of his new and more powerful role as chief of the fire staff and inspector in chief. He oversaw standardisation of equipment and procedures and established a national training college. At its peak he led 370,000 staff, including 80,000 women – he had become a convert to the idea of employing them.

The FBU, which welcomed nationalisation, called for it to be also used as an opportunity to introduce a national minimum rate of pay, and to settle many of the firefighters' grievances. The month after nationalisation, the union launched its Fireman's Charter. It wanted an end to such things as stoppages from a man's basic pay as a means of punishment.

Nationalisation, the FBU said, should be a chance to correct the appalling system which had left Harry Errington without a job because it took him more than thirteen weeks to recover from injuries sustained while fighting a fire. The Charter demanded full pay while an auxiliary firefighter was sick or injured – exactly what he would get if he was a regular. In December 1941, the FBU secured a deal which gave up to 26 weeks injury pay, and higher wages.

Herbert Morrison talked of using nationalisation as a pretext for outlawing the independent fire service trade union, saying that firefighters would now be Crown servants. But he was persuaded that it was not a good time to get into a fight with the firefighters' trade union, and the next year he formally recognised the FBU as the negotiator for firefighters. Nonetheless, he continued to be interested in ways in which he might set up a tame competitor; and he actually set up a union for officers, with none other than Firebrace as its president, which siphoned senior men away from the FBU.

But the 1943 FBU conference, the union's 25th anniversary, offered a chance to bury the hatchet, and Morrison, Firebrace and Dixon all nailed smiles to their faces and turned up for the Jubilee Dinner of sausage tart and two veg and treacle pudding. "Dixon surprised us with all his charm, and Commander Firebrace genuinely tried to unbend before a couple of hundred of his rank and file," reported John Horner.

One immediate benefit was the production of the first comprehensive *Manual of Firemanship*. Seven volumes were published in 1942, along with the NFS *Drill Book*. Gradually the work was being standardised throughout the country. "This book is one of the results of nationalising the fire service," noted the FBU magazine *The Firefighter*. "Its scope as well as its cost would have put it beyond the capacity of any local authority, while no body outside the Service itself could have gained access to all the necessary material."

So nationalisation got rid of many of the idiocies of the old system – but, it has to be said, it created a few idiocies of its own. "Imagine," writes Frederick Radford,

> *the effect during an all-out war effort of an instruction that firemen in the NFS would salute their officers above the level of company officer! Imagine with what incredulity firemen heard that a Labour minister was responsible for this order, and had detailed that the salute should be "naval style" and that women were to salute men officers, but that men would not salute women officers.*

The name on the order was Morrison's, but surely the mind behind it must have been that of Commander Firebrace.

And nationalisation did not get rid of the long-standing grievance over hours. The 48 hours on – 24 hours off duty system was still in operation, and 112 hours was too long for anyone to be asked to work in a week. There were ballots on strike action, and the FBU leadership was in a quandary – they knew the members were right, but did not want to have a dispute that might weaken the war effort. It was not until 1944 that the government finally agreed to what the firefighters wanted – an 84-hour system.

*

Lessons from the morale-sapping boredom of the pre-Blitz days were also learned. The Workers Educational Association became more active in the north of England in arranging classes for firefighters, while in the south, in 1941, London's City Literary Institute took a hand in developing the spontaneous discussion groups that had grown up. The Institute trained 80 firefighters to lead discussions on such subjects as Russia, the colonies, and what to do with Germany when the war was won.

The next year the government published the Beveridge Report, calling for an attack on the five giants – Want, Ignorance, Disease, Squalor and Idleness – and a welfare system that looked after citizens from the cradle to the grave. This swiftly became one of the most popular subjects for discussion in these groups. That is not surprising when you remember that the report sold out its huge print run on the first day it was published, which sounds incredible today – a dry government report, with not a picture in sight, becoming a best seller. People could at last see an end to the war, and hope realistically for victory, and the enthusiasm for Beveridge was a way of giving notice to the government that after all the sacrifices that victory was going to demand, they were not going to go back to the same old unfair society, and to the cruelties of thirties poverty and unemployment.

The organisation of education in the fire service became remarkably sophisticated as the war progressed. There was an organiser in each fire force, and two men in each force were released from other duties to become regional organisers. The FBU frequently had to step in to protect their facility time, so that they could do their work of ensuring that the discussion leaders knew what they were doing and how to get the best out of the discussions. They arranged weekly lectures for the leaders on topics that might be suitable for discussion.

This venture proved unexpectedly popular in fire stations, where it had to compete with such activities as playing cards, snooker or billiards. The discussion groups led to a series of lectures which many firefighters attended enthusiastically.

Among the new recruits to the AFS in the late summer of 1941 was the 32-year-old public-school-educated poet Stephen Spender. He became one of the first organisers of discussion groups. He expected to find the atmosphere a little like his old school, but it was much kinder than that: AFS men looked after each other, and were "tactful, and light and easy". The three-week training came as a shock – climbing very long ladders, and finding that all the equipment he had to learn to handle was "wet and cold and intractable and heavy". Then he was sent to a fire station which was like many of the time:

We were housed in three huts. One was a dormitory, one a recreation room, and also, in part, a dormitory, and the third one a kitchen and mess room. There were always 20 men on duty in this station, and usually ten or more of them were in the recreation room, unless we were doing a drill or cleaning the station.

But he found that to some extent the station, like many others, was living on memories of the Blitz. The men who had been serving then would tell endless stories about it. Compared with those heady days of fighting furious fires night after night, they were starting to find fire station life boring.

In early 1942 Spender was transferred to a very different station. "Life was much easier here," he wrote. "There were many rooms – it was an old LCC school building – and one could quite often get away alone. There was less brass polishing to do, and more men to co-operate in keeping the station floors scrubbed."

As an organiser of discussion groups, he got to see several other stations too. "In some the men were keen and interested, not just in discussions but in everything they did. The keenness of men at such stations to learn and discuss and to improve conditions for themselves and their children was far greater than that of most undergraduates at Oxford or Cambridge." He wrote of finding in all fire stations "the seriousness of people who have lived close to death, and who have always seen destructive and impoverishing conditions around them".

There was another use for spare time. At station after station, firefighters started to think that, while there was less firefighting to do than they were used to, they could be putting their time and skills into other ways of advancing the war effort. Slowly the idea started to catch on nationally, and the NFS and the FBU began to get involved.

Forty-five stations worked for one firm producing components for Lancaster bombers. In another station they produced parts for aircraft petrol tanks. At Southampton a union employment bureau was set up to find work for firefighters on their leave days. In one Preston fire station, 400 brass cuttings for Spitfires were made on off-duty days. Some firefighters built farmhouses, others felled timber, yet others kept pigs, rabbits and poultry. By March 1943, 700,000 man hours had been worked in London alone.

The only person who seemed unimpressed was the FBU's old adversary Herbert Morrison. When *The Times* congratulated the firefighters and criticised the government for its lack of interest in the scheme, Morrison told the House of Commons that the FBU had achieved "the amazing success . . . of getting on the leader page of *The Times*. It was first-class propaganda."

The lull also provided a chance for firefighters to take stock of their changed situation, and the biggest and most startling change was not that there were now many times as many firefighters, but that a substantial number of them were now women. There had not been a single woman in the fire service before the war, and their presence would have seemed unthinkable. By 1943 there were 4,800. The FBU had women organisers and a woman on its national executive, for the first time ever, and was demanding the same conditions for them as for the men.

The scene for one of the FBU's earliest victories for its new women members was the Royal College of Surgeons, now housing a makeshift dormitory for AFS women. Horner told the tale:

> *The aged gorilla at the zoo, Mog, whose antics had made him a favourite with visitors, passed away. The zoo presented Mog to the Royal College for the study of comparative anatomy. There had been no room for Mog in the removal vans when the College was evacuated and he was left behind in one of the corridors, supine with arms above his head, in a tank of formaldehyde.*

The tank was in the corridor that led to the women's dormitory, and they asked Horner to intervene with the authorities and get it taken away. He succeeded.

*

In late 1941, a number of popular musicians initiated the London Fire Forces Dance Orchestra. One of the alleviations of wartime life was the music of the various service dancebands, such as the Royal Air Force Dance Orchestra, also known as the Squadronaires, the No. 1 Balloon Centre Dance Orchestra, alias the Skyrockets, and the Royal Army Ordnance Corps Dance Orchestra, alias the Blue Rockets. All these units drew on the large pool of professional dance band musicians who, before the war, had worked under bandleaders like Billy Cotton and Lew Stone. The London Fire Forces Dance Orchestra, according to Harry Francis, its drummer,

> *was never an official Service band but one given facilities to rehearse and to broadcast whilst on duty, but there was an additional and important reason for its existence. During the air raids upon London and elsewhere of 1940 and 1941, we had lost many hundreds of comrades . . . with the result that there were many widows and orphans for whom official compensation was inadequate. The orchestra was therefore formed on the basis that it would give performances, without fee, to raise money for the Fire Service Benevolent Fund . . . Our library contained the work of some of Britain's finest arrangers of those years, such as George Evans, Phil Cardew and Sid Phillips.*

A war-time journalist once described us as being "easily our second-best National Service band" and added that this was the equivalent to saying "second-best in the whole country" but, although this was certainly meant as a compliment, I think that, like myself, few members of the orchestra agreed with him. Restricting the matter to Service bands, I think I would have put the Squadronaires first, with the Skyrockets and Blue Rockets in second and third places in either order and our orchestra running fourth . . .

Eric Midwinter recalls that their signature tune was "I Don't Want to Set the World on Fire".

The fire service could also boast silver and brass bands in various parts of the country. In Durham, for example, the Easington Colliery Youth Band became, for the duration of the war, a National Fire Service Band – a title it shared with other such aggregations in Portsmouth, Southend and elsewhere.

*

One of the lessons learned from the Blitz was that prompt reporting of fires was a key to success. Failure to report fires in the City of London had led to unnecessary losses. Fires were often reported inaccurately, and communications had been known to become overloaded with the reporting of one fire, while others were not reported at all. The control centres sometimes received up to sixty reports of the same fire. So an elaborate fire guard system was created. "British citizens," stated *Front Line*, "were forced to the realisation that civil defence was everybody's business, and that the answer to concentrated fire attack required the watchful eyes and quick hands of roof and street patrols twenty times more numerous than fire brigades could ever be."

Roy Sargeant remembers how

fire guard groups were formed in local neighbourhoods. This was a first-strike response to incendiary bombs and small fires. They were issued with large grey, net-curtained steel helmets [and] armbands [with] "FW" and worked on a roster system.

I would come home at maybe 0200 hrs after Wailing Winnie [a siren] had declared All Clear with her continuous unbroken whine, and find my Mum with helmet by her side, sitting in the chair, wide awake, having a cup of tea. Despite my efforts to get her to go to bed Mum would say, "No, I'm on duty," so I'd point out the "all clear" had gone but Mum would still stay on watch either indoors or in the street, until her duty hours were over.

Eric Midwinter remembers the fire guarding of his street in Sale.

Fire guards were formed in each road. My mother was a fire guard, a member of a little group in the road where I lived who were supposed to run out with a stirrup pump and put out fires, with their white helmets on . . . They were trained by firemen to use the stirrup pump. That was a foot pump where a pipe went into a bucket of water and you treadled your foot up and down. Men came from the council and they put "SP" on either side of our front door, for stirrup pump. We children were very proud of this.

And then for some reason it was decided that a stirrup pump shouldn't be held in the house of a fireman. So it had to be removed and it was taken across the road to Mr Bird at number 13. Round came the man and scrubbed away the "SP" – leaving for many years this faint "S" and "P" – and to our chagrin put this very white "S" and "P" on each side of Mr Bird's front door.

At the start of the war, leaflets distributed to every household had warned: "IF YOU THROW A BUCKET OF WATER ON A BURNING INCENDIARY BOMB, IT WILL EXPLODE AND THROW BURNING FRAGMENTS IN ALL DIRECTIONS." Consequently fire guards were provided with

Borough of Sale

FIRE
GUARD
CERTIFICATE

BOROUGH OF SALE

5080

This is to Certify that

Mrs Midwinter

of 1b Lynwood Grove

Sale

is recognised by the SALE BOROUGH
COUNCIL as a Fire Guard, and
possesses the powers of entry and of
taking steps for extinguishing fire or for
protecting property, or rescuing persons
or property, from fire, which are conferred
by the Fire Precautions (Access to
Premises) No. 2 Order, 1941.

Any individual who has been
recognised by the Sale Borough
Council as a Fire Guard is
eligible for benefit in respect of
"War Service Injuries" under
the Personal Injuries (Civilians)
Scheme, 1941.

Town Clerk.

Town Hall, Sale.

.................................194

Member's
Signature E. Midwinter

Mrs Midwinter's fire guard's certificate © Eric Midwinter

buckets of sand, which were more effective, as were sandbags. The fire guard who moved promptly as soon as the incendiaries landed, and was equipped with thick gloves or a pair of fire-tongs, could pick one up, drop it in a bucket and remove it to a safe place – but this became much riskier when the Germans began fitting a proportion of their incendiaries with explosive charges.

Four million pounds were allocated by the government to ensure better supplies of water for the fire brigade, including borehole pumps on river barges and wharves in the Thames, so that water could be pumped to street level even when the river was at low tide. Pay for a NFS firefighter went up to £3 10s. (£3.50) a week. AFS men began to take on work in the fire service for which their pre-war occupations as plumbers, fitters, bricklayers and in similar trades fitted them – after negotiation with the trade unions to agree appropriate rates of pay for the use of such skills.

Chapter 11
1 February 1942–11 June 1944
The Baedeker Blitz

While the Germans never again managed to bomb Britain on such a large scale, they carried out smaller attacks throughout the war, taking the civilian death toll to 51,509 from bombing. In 1942–43 there were only 23 air raids over London, and the LFS was able to contain them effectively. They were even able to send men and machines out of London to help local brigades deal with what became known as the Baedeker Blitz.

The Baedeker Blitz was a series of raids conducted in 1942 as reprisals for the RAF bombing of the German city of Lübeck. These raids, mostly between February and May, targeted historic cities of no military or strategic importance: in particular, Exeter, Bath, Norwich, York and Canterbury, which was bombed three times in a week in May–June. The raids derived their name from the famous Baedeker guidebooks, which were to tourists in the 1930s what Lonely Planet is to us. It's said that Hitler and Göring chose their targets by consulting a copy of Karl Baedeker's *Great Britain: Handbook for Travellers*. Churches and other public buildings of interest were often the targets of these retaliatory raids (*Vergeltungsangriffe*) in an attempt to break civilian morale. But although some well-known landmarks such as the Guildhall in York and the Assembly Rooms in Bath were seriously hit, historic churches such as Exeter and Norwich Cathedrals escaped serious damage. There was, however, considerable loss of life – over 1,600 fatalities – and some 50,000 houses were destroyed. The Luftwaffe suffered considerable losses, outweighing the propaganda value of the relatively minor damage it inflicted.

Relatively minor compared with what had been done to

London, that is: but it was devastating to the inhabitants of these cities. In Exeter, a raid on 4 May killed 156 people and injured 563, many seriously. Among the buildings lost were Deller's Café, Bedford Circus, St Lawrence Church, the Lower Market, the Globe Hotel, Dix's Field, the College of the Vicars Choral and the City Library, which had been the local fire service's Control Centre. Much of the High Street was destroyed and so were many houses in the Newtown district.

Norwich endured raids in April and May. Richard "Dick" Waller, a firefighter based at the Surrey Street sub-station, remembered one of those nights for the *Norwich Evening News* in 2007.

> *We went to two fires that night and after we had put out the first we were told to go to the Trafford Arms which had been hit. It was well alight when we got there – it was just razed to the*

The East Coast Main Line and its centres of engineering like York were prime targets between 1940 and 1943. The Sir Ralph Wedgwood A4 class 4-6-2 steam locomotive no 4469 lies damaged in the wreckage of York North locomotive depot, following an air raid on 29 April 1942. © Science & Society Picture Library

ground. We were there for two hours, but they were still bombing while we were trying to put the fire out.

Incendiaries were falling all around them. "Some of the incendiaries were very phosphorous, meaning if there was a bit that dropped on your hand it would go right through, so we tried to avoid those ones." When they finally doused the flames, the firefighters checked the inside of the pub for survivors who might have sheltered under stairs or in the cellar.

If there was anyone alive or dead that's where they'd be. We went in and there was no one there – everyone had got away, but I saw a bundle of clothes in the corner and thought we had found a body. But it wasn't a body – it was the pub's till.

He recalled, too, the camaraderie of the firefighters.

We did worry about each other, we knew everything about each

What the centre of Exeter looked like a few months after it was bombed.
Photo courtesy of exetermemories.co.uk

other – we used to get so close to each other because we relied on each other for our own lives. When we finished we would walk home with craters everywhere and turn a corner and wonder if our houses and families would still be there.

First World War veteran Harry Patch, whom we met in Chapter 1, was a firefighter during the Baedeker raids. He had returned to his trade as a plumber, but joined the AFS in 1938 at the time of the Munich crisis and was based at a fire station in his village of Combe Down, near Bath. Bath was not well defended or prepared for the Baedeker raids. Patch fought fires in Wells Way, which led from Combe Down to Bath, and remembered all his life a bomber flying low over his crew and dropping flares for illumination, then opening fire on the firefighters with a machine gun. He dealt with four major raids and later told his biographer: "Did the bombs remind me of Ypres? Of course they did. I was going through it again, and it was tough."

Alfred West, a Stoke-on-Trent firefighter who had

This paste up of group photographs of National Fire Service personnel appeared in The Story of the Exeter Blitz, *which was published during the war. Fire crews from Plymouth, Taunton, Torquay and even one from Reading drove to Exeter, on poorly maintained roads, to assist the local fire service to douse the flames.* Photo courtesy of exetermemories.co.uk

volunteered to go to the south-east where raids were more likely, was injured in one of the raids on Maidstone, when a sheet of glass pierced his helmet. He returned home, his face a mess, and his wife picked him up at Stoke station. Their son Linden West remembers her telling the story of his ghastly appearance and how she said to her husband, "You make me so ashamed!"

As the Baedeker Blitz was under way, the Allies were bombing Germany – and with much greater effect. Firefighters had been consulted by the RAF Bomber Command, who asked them, "In the light of your experience during enemy air attacks, what change of tactics by the Luftwaffe would cause you most concern?" Firefighters knew the answer. The concentration of a heavy attack into a very short space of time could swamp fire service resources. Fires would burn unhindered because there was no one left to put them out, and

The centre of Bath, the view from Julian Road showing bomb damage to St Andrews Church (left), August 1940. © Robert Hunt Library / Mary Evans

there would not be enough water.

So when 1,000 bombers attacked Cologne in May 1942, 1,500 tons of high explosive were dropped on the city in the space of an hour and a half, and fire services were overwhelmed.

*

During 1942 Britain's fire services were augmented by firefighters from Canada. The idea had been conceived by the Canadian Prime Minister, William L. Mackenzie King, during a visit to London in summer 1941, when he saw the consequences of the Blitz. As soon as the first news of the Blitz reached Canada, many full-time firemen and civilians expressed a desire to cross the Atlantic and help the British fire service. During the early months of 1942, Canada's fire chiefs assembled more than 400 of these volunteers, drawn from all nine provinces and over a hundred municipalities. In March they were given the title of the Corps of Canadian Fire Fighters, and on 24 June the first contingent arrived in Britain. By December, the CCFF numbered 422 men.

Some training was necessary; the trailer pump, for instance, an integral part of British firefighting, was unfamiliar to Canadians. After that, they joined the NFS at fire stations in London, Bristol, Plymouth, Portsmouth and Southampton. One of them, Sam Posten from Regina, Saskatchewan, stationed in Plymouth in 1942–44, took many graphic photographs of the damage done by German bombs in Plymouth and London. Three Canadian volunteers died while serving in England.

*

Two 1943 films celebrated the contribution of the wartime fire services. Many veterans and their families fondly remember *The Bells Go Down*, an Ealing Studios feature that tells the story of three young men who join the AFS at the beginning of the war and have their mettle tested in the Blitz. A sub-plot touches on the rivalry between the part-time AFS and the full-time LFB (although, by the time the film was made, the two brigades had been brought together as the NFS).

Several distinguished cinematic names of the period were involved in the production: director Basil Dearden, producer Michael Balcon and, among the cast, Tommy Trinder, James Mason, Mervyn Johns, William Hartnell and Finley Currie.

The cinema historian Mark Duguid writes,

> The Bells Go Down *was among the last of Ealing's wartime films to take the conflict as its subject – by 1943, the war was already on the turn, and the Blitz could be remembered with a tinge of pride, as an early challenge that was seen through with courage and fortitude. So the tone of the film, even though it allows for tragedy, is largely celebratory: a testament to the bravery and endurance of volunteer firefighters . . . it is about teamwork and comradeship. But* The Bells Go Down *emphasises solidarity not just among the fire crews, but in the community at large. Petty differences are set aside as neighbours pull together to support one another through difficult times. Similarly, class divisions are forgotten – the fire chief is a bluff Scot . . . and the one upper-middle-class fireman . . . is not an officer but simply one of the crew.*
>
> *The film was released in May 1943, just a few weeks after Humphrey Jennings' semi-documentary about the AFS,* Fires Were Started, *and seemed to confirm that Ealing's filmmakers were keeping a close eye on their counterparts in the documentary film movement, which may be why an early title considered for* Fires Were Started *was* The Bells Went Down.

Frank Harbud remembers his father Charles, who was based at Parnell Road Fire Station in east London, taking him to see the film.

*It was made on his ground. He made me see a matinee show, then
took me up west for a lobster dinner and I slept that night on his
canvas cot in his office at the station. Every time I woke up he
was hard at work at his desk seeing to the admin and running of
the stations he controlled. If ever a bloke burnt the candle at both
ends – he did.*

Clive Skippins' father was stationed in Walthamstow. "He
took me to see the film in 1944 or 1945," he writes. "This has
remained one of my favourite films of all time and is probably
why I spent 36 years of my life in the Fire Brigade."

There is a documentary element in *The Bells Go Down*: a
good deal of the footage had been shot at real fires during the
air raids of the previous two years. But a more consistently
realistic portrayal of the firefighter's job in the Blitz is offered
by Humphrey Jennings's *Fires Were Started*, a feature-length
dramatised documentary in which all the fire station
personnel, from the men on the "heavy units" and trailer
pumps to the "girls" in the control room, are played by men
and women who actually did those jobs. One of the firefighters
who appeared in the film, William Sansom, said later: "The
way the firemen behaved in the film was absolutely authentic,
absolutely the way it was. There was only one significant
difference: in real life, everybody was swearing all the time."
Another of the firefighters, Fred Griffiths, said: "I tell you this:
as you see it depicted here, so it was – and sometimes far
worse."

In a warm appreciation of Jennings's work, published in
Sight & Sound in 1954, the director Lindsay Anderson gave a
detailed description of what he regarded as one of the
documentary-maker's masterpieces.

*[It] is a story of one particular unit of the National Fire Service
during one particular day and night in the middle of the London
Blitz: in the morning the men leave their homes and civil*

occupations, their taxi-cabs, newspaper shops, advertising agencies, to start their tour of duty; a new recruit arrives and is shown the ropes; warning comes in that a heavy attack is expected; night falls and the alarms begin to wail; the unit is called out to action at a riverside warehouse, where fire threatens an ammunition ship drawn up at the wharf . . .

(The area that sub-station 14Y is responsible for in the film is in Limehouse, around Trinidad Street and Alderman's Wharf. The fire that "Heavy Unit 1" is called out to deal with was staged in an already burned-out warehouse, but the long shots of blazing riverside buildings are in fact footage of the Tate & Lyle factory at Silvertown, which was bombed on the first night of the Blitz.)

. . . the fire is mastered; a man is lost; the ship sails with the morning tide. In outline it is the simplest of pictures; in treatment it is of the greatest subtlety, richly poetic in feeling, intense with tenderness and admiration for the unassuming heroes whom it honours . . . No other British film made during the war, documentary or feature, achieved such a continuous and poignant truthfulness, or treated the subject of men at war with such a sense of its incidental glories and its essential tragedy.

*

After the Baedeker raids, there were occasional attacks on the rest of the country during 1943, including a very heavy bombardment of Grimsby and Cleethorpes in the early hours of 14 June which killed 101 people and injured another 300. Fire brigades came in from Mansfield, Lincoln and Leicester, but they were severely hampered by low water pressure. The two towns were hit again a month later, this time killing 65 and injuring 173.

In November 1943, Göring ordered the Luftwaffe to resume mass bomber attacks against southern England. During

December 1943 and early January 1944, the Luftwaffe gathered some 515 aircraft of widely differing types on French airfields. On 21 January 1944, these aircraft made the first mass attack on London since 1941. A force of 447 bombers was sent out, including Junkers, Dorniers, Messerschmitts and the new Heinkels. The bomber crews generally lacked night flying experience, and the aircraft types had very different standards of performance. The Germans were forced to use pathfinder aircraft to mark targets. The raid was a failure: only 32 of the 282 bombs dropped fell on London that night.

The raids continued for the next three months, to little effect. The Germans lost 329 aircraft – which were then not available to defend against the Allied invasion of continental Europe. By 1 July 1944, Germany had only 90 bombers and 70 fighters available in Western Europe.

These raids gave Stephen Spender his first taste of real firefighting. He found the waiting worse than the firefighting itself. Once he heard the bell and rushed into the road to jump on the fire appliance, there was a sense of elation.

"What I remember better than flames pouring out of windows," he wrote, "is the hundreds of incendiaries lying in the Kensington streets on our way to the fire, like many workmen's hurricane lamps."

*

While Spender was getting his first taste of firefighting, 20,000 firefighters were being moved southwards in the greatest secrecy, to live in makeshift fire stations along the south coast, in preparation for D-Day. The Home Office and the FBU worked closely together over this, and the FBU took charge of all the welfare arrangements, working freely in militarised areas from which the civilian population had been evacuated.

Forty years later, another Prime Minister would say that membership of a trade union was incompatible with national security, and ban union membership at GCHQ; but when the nation's existence really was at stake, in 1944, Prime Minister Winston Churchill cheerfully made an ally of the radical left-wing FBU, and shared military secrets with it. While the Mulberry Harbours – those extraordinary constructions that were eventually towed overnight to Normandy and used to enable troops to disembark on the beaches – were being secretly built in British ports, firefighters were there every day keeping the hard area free of mud and silt.

The Mulberry Harbours were eventually towed to the Normandy beaches on 6 June 1944. The invasion of Hitler's Europe, and the last act of the war, had begun. But Hitler had one more aerial weapon to throw at Britain before he was finished. Six days after D-Day came the first of the flying bombs.

Chapter 12

12 June 1944–VE Day
The flying bombs

In the summer of 1944, Roy Sargeant, by then training as a sick bay attendant in the Royal Navy, was firewatching on the roof of a hospital at Gillingham in Kent.

One night there was what appeared to be a normal heavy air raid with plenty of anti-aircraft fire and, when we went out on to the balcony, we saw aircraft apparently with tails alight scudding across the sky. A flame was shooting out of the rear of these aircraft as though they were on fire.

Sargeant had sighted some of the first V-1s: pilotless flying bombs launched at London and south-east England from German bases in Europe. Each V-1 carried 150kg of high explosive at 350mph. They were known as buzz-bombs or "doodlebugs". They first attacked London on 12 June 1944, and, in all, 638 came down in June alone, although 30 a day were intercepted and destroyed by RAF fighters. Eventually, over 4,000 were destroyed by the RAF, the Army's Anti-Aircraft Command, the Royal Navy and barrage balloons. The worst day was 3 August, when the fire service attended 97 doodlebug incidents in and around London.

Planes were seldom effective against them, and they kept the NFS constantly busy. Firefighters went to wherever they fell, even when they did not cause fire, because they were still needed to help get the casualties away and start to repair buildings – the bombs would bring down buildings over a wide area. Some of them did start huge fires, as when one fell on a Thameside candle works that was full of cans of paraffin wax, fuel oil and turpentine. Firefighters worked knee-deep in

oil and molten wax, and it was a miracle that none of them was killed. Another V-1 hit a Kensington gasometer, another a building full of tar containers. Firefighters dumped sacks of sand in the path of a river of flowing, burning tar.

W. J. Burningham, an NFS Mobilising Officer, wrote an account of a flying-bomb incident he witnessed.

One of our dock fire stations had been badly damaged by a near miss. I was ordered to visit the Albert Dock and decide where we could set up a temporary control and appliance bay for the dock fire appliances.

I was proceeding on my way in my fire car and when about a quarter of a mile from the dock entrance I heard the sound of a flying bomb approaching. The clouds were very low and it was raining quite hard, so that it was not possible to see the bomb in the sky.

Smoke and flames rising from a burning garage at Elmers End, near Beckenham in Kent, which received a direct hit from a doodlebug, 26 September 1944.
© Science & Society Picture Library

Suddenly the noise of the engine of the bomb ceased, which I knew was a sign that it had used up all the fuel and was about to fall.

Looking in front of me I saw it fall through the clouds and into the dock, exploding in the water less than 200 yards from a ship that had come from America filled with explosives.

The ship's deck houses were badly damaged, but it was most fortunate that the ship was not hit, otherwise a lot of the surrounding transit sheds, which were stocked with shells and mines, would have also exploded.

Firefighters saw terrible things in the wake of the flying bombs. "I've never seen such sights, either on battlefields or during the worst part of the previous Blitzes," said one. Here is part of a report by Chief Regional Fire Officer Delve:

Company officer Dew forced an entry into a basement where he found two women. He enlarged the hole to enable a doctor to join him and it was then found that one woman was dead and would have to be moved to free the other woman. With the help of the Heavy Rescue Service the second woman was rescued alive. Company officer Dew was in the basement for nearly two hours with a quantity of debris above him and his good work did much towards saving the woman from asphyxiation.

Firefighters used 100-foot turntable ladders to bring people down from the fourth, fifth or sixth floors of buildings. Occasionally a ladder would be employed as a crane on which to strap a badly wounded person.

The most harrowing task was rescuing children. On one occasion firefighters worked for three hours in a space about three feet high and managed to free four children, but two were already dead and the other two died later in hospital. A firefighter named Phillips, based in Poplar, remembered

*a rescue squad releasing a small child that had been buried in the
collapse of a house. The child had been buried standing up, and
was obviously dead. The rescue party were digging away, and
they had just uncovered his head and shoulders. It was a terrible
morning – rain was pelting down – and all I can see now is that
child's head and shoulders, standing above the debris, white-faced
and clean where the rain had splashed and washed the face of the
child.*

Most firefighters kept such memories to themselves, and
even years later, after the war, were still reluctant to talk about
them. Rosemary King, whose father Richard was an AFS
firefighter in Birmingham, says:

*Seeing old newsreels and films of those days, we find it hard to
believe that ordinary people, like our father, could show such
courage and tenacity. He must have encountered horrific things,
but rarely spoke about those traumatic years and when he did, it
was in a matter-of-fact, self-deprecating way.*

Harold Newell told his daughter Lyn, "Once you've pulled
a dead child out of a burning building, you never forget it."

*

The V-1 was succeeded by the V-2, first used against London
on 8 September 1944. This was a deadlier weapon, a rocket
carrying a ton of high explosive, and, unlike the V-1, it came
silently, launched from Holland and arriving in Britain less
than five minutes later. Roy Sargeant recalls that there was "no
swish, no whistle, no warning. Just a massive explosion – then
we would see the vapour trail ascending into the sky, after the
rocket had already hit and exploded. These were the worst air
raid weapons. Completely indiscriminate, death with no
warning at all."

A V-2 often killed 100 or more people. There were fifteen

Firefighters rescue people, Queen Victoria Street, London, 1939
© London Fire Brigade

V-2s in September, 25 in October, rising to a peak of 116 in February 1945 and 115 in March. In all, 1,115 V-2s were fired at the United Kingdom, killing an estimated 2,754 people in London and injuring another 6,523.

A total of 9,251 V-1s were fired at Britain, the vast majority aimed at London; 2,515 reached the city, killing 6,184 civilians and injuring 17,981. Altogether, the V weapons killed 8,938 civilians in London and the south-east.

The last bomb to fall on London was a V-2 on Tuesday, 27 March 1945. It was one of the most dreadful, destroying two of the three blocks of council flats known as Hughes Mansions in Stepney, east London (now part of Tower Hamlets). The flats were built in 1926 and named after Thomas Hughes, the author of *Tom Brown's Schooldays*. For Britain's Jewish community, it seemed like Hitler's last attempt to get his hands on them: a large number of Jews lived in Hughes Mansions,

A firefighter and his family outside Buckingham Palace after a medals presentation on Victory in Europe Day. Photograph by Frederic G Roper.
© Science & Society Picture Library

and of the 134 people killed, 120 were Jews. Whole families were killed. Bernard Cohen, a young soldier fighting in Malta, was given the dreadful news that his parents and three sisters had all perished in Hughes Mansions.

At the end of the Blitz an estimated 16,000 had been killed and 180,000 injured. In London alone, 320 men and woman of the fire service had been killed in action, and more than 3,000 injured. In addition, 662 of the 875 fire stations were damaged.

But it was the warmest March for years, and an end to the war was in sight. Germany no longer had the capacity to bomb Britain. There was a stand-down parade for AFS people on 18 March 1945, and that evening they were invited to a "Cavalcade of Stars" at the Albert Hall, with, among others, Vera Lynn and George Robey.

Chapter 13
After the war

After VE (Victory in Europe) Day on 8 May 1945, several of the auxiliaries found firefighting a rewarding way to spend their working lives, and stayed in the service. Others went back to their old jobs or, in the case of part-time auxiliaries, stayed in them. The FBU, whose membership had been 60–70,000, shrank – at one point to just 12,000, though it picked up again.

AFS man Charles Cunnington remained a firefighter until the early 1960s; his son John followed him into the service. John's brother, son, son-in-law and grandson have all also spent time in the fire service – making five generations of Cunningtons connected with firefighting, for Charles's father Tom had been a volunteer firefighter in the days of horse-drawn engines.

Harold Midwinter also stayed in the fire service. Towards the end of the war he had volunteered to go into the maintenance side, based in Manchester, where he worked until 1947. Then, his son Eric remembers, "He saw an opportunity that seemed very good: he became the fireman at the Palace Theatre, Manchester, from six in the evening till six in the morning. And I saw all the shows. He did that for two or three years, then came out of firefighting altogether." It was a chance for young Eric to get to know and love the music hall and variety theatre, a passion that has remained with him throughout his life, and about which he has written several books, including *The People's Jesters – Twentieth Century British Comedians* and *I Say I Say I Say – The Double Act Story*.

Harry Errington resumed his trade as a tailor, but was always a welcome visitor at his local fire station in Soho, rebuilt after the bomb that destroyed it, where he was given a ninetieth birthday party in 2000. He died in December 2004, aged ninety-four.

Hubert Idle went back to being a research chemist. His discharge certificate and that of his wife Joyce, both dated 31 January 1946, give as the reason "reduction in part time establishment". But the bond between him and the fire service remained strong all his life. "When an old comrade died he'd keep in touch with his wife, when she died he'd keep in touch with the children," says his son Angus. Some of his comrades had been killed when a bomb hit the fire station where he had been based, in Wandsworth, and they were in his mind all his life. When he died in 2009, at the age of 104, Wandsworth Fire Station turned out for his funeral.

Harold Newell returned to his trade as a furrier in Bounds Green, but he always retained his links with the NFS. His daughter Lyn Nixon remembers the NFS club in White Hart Lane – the camaraderie, the billiards, going to dances there in the 1950s – as well as outings on the *Royal Sovereign*, a pleasure boat moored at Tower Bridge, down to Southend, "the boat lit up and everybody dancing".

> *He did enjoy his time in the fire service . . . but he felt quite sad because there was no memorial to the firemen in his lifetime, and they didn't march at the Cenotaph after the war, and I think he felt that that wasn't right. And I know a lot of the people that he was at the fire station with, they thought that too . . . I remember he used to watch the Remembrance Day service and he always said, "And where are we? And where are we? It's wrong. We lost so many men, and we did a lot of good, we saved a lot of people" . . . He always felt that the work the fire service did was never properly understood or remembered.*

He had no pension. "He never got a penny," says Lyn. She thinks this was generally true of the AFS.

Some regular firefighters were also treated shabbily over their pensions from their wartime service. Alfred West lost out on his pension because of his decision to volunteer to go south

to fight the Blitz. West had joined the Stoke-on-Trent fire brigade just before the war, and when the Blitz was at its height, volunteered for service in the south east. He spent several years in Kent, where he witnessed the June 1942 Baedeker raids on Canterbury and, as described in Chapter 11, was injured by flying glass in a factory fire in Maidstone. But this period away from his original post in Stoke-on-Trent, as his son, Professor Linden West, explains,

Alfred West © Linden West

meant a disruption of his pension. And you would have thought that the consequences of volunteering would not have led to that. Because it wasn't a consecutive service with the same employer, technically, he lost those years from his pension . . . The FBU fought a long, long campaign over this. It dragged on and on and on, and there was obfuscation piled upon obfuscation. Unfortunately it wasn't until some years after he died that the thing was put to rights. This was a source of great family pain.

*

Almost exactly two months after VE Day, on 5 July 1945, Clement Attlee became Prime Minister at the head of the nation's first Labour government with an overall majority. Labour was returned with 393 seats. The Conservatives had 213, the Liberals 12. With two Communists, one Commonwealth Party member and 19 others, Labour had an overall majority of 146.

A secret last-minute attempted parliamentary coup by the FBU's old nemesis Herbert Morrison, who felt he should be Prime Minister rather than Attlee, failed.

John Horner could easily have been among the Labour MPs – and his old opponent Herbert Morrison seems to have hoped he would be – but he decided, first, that he wanted to stay with the union for what promised to be some of its most difficult years; and, secondly, that it was time to join the Communist Party, with which he had always had some sympathy.

Despite universal threadbare poverty, rationing and Britain's wrecked economy, it seemed a moment of hope the like of which no one could remember. The new government was committed to the first serious assault on the social ills identified by Lord Beveridge in his 1942 report. The main instrument was the welfare state. Beveridge had designed the building; Attlee and his ministers set themselves the task of erecting it, to a deadline of 5 July 1948.

The government was also committed to narrowing the

enormous gap between the very rich and the very poor, and to a huge programme of nationalisation. Perhaps best of all, the NFS was no longer responsible to Herbert Morrison. He had been placed in charge of the nationalisation programme, and the new Home Secretary was Chuter Ede, with whom FBU leaders met in October. They told him of their troubled relationship with his predecessor, and they protested about the programme of "reassessment", which meant that men who had fought every type of fire during the war might be told that they were no longer considered competent in peacetime, perhaps because they were an inch or two below the minimum height.

One auxiliary who joined in August 1938 and became full-time in September 1939, who fought fires in Docklands and all over the country, then went back to London to fight the doodlebugs, and asked to stay on, was told, "The chief regional fire officer regrets you do not possess the necessary qualifications and experience."

Chuter Ede agreed to change the system so that men could stay unless they were unfit or a danger to others.

*

Firefighters might have been forgiven for hoping that, in the new political order, when the mines, railways and banks were all being nationalised, the nationalisation of the fire service would become permanent. They were to be disappointed. On 1 February 1947, the Fire Brigades were split and once again organised on a regional basis. The NFS was transferred back to local authorities.

Commander Firebrace, who had become Sir Aylmer Firebrace in 1945, retired three days before the fire service reverted to local authority control. He always claimed thereafter that during the six years of its existence the NFS had made progress that otherwise would have taken many years to accomplish.

But it was not a full return to the pre-war system. Lessons had been learned, and new safeguards were built in. The number of peacetime brigades was drastically reduced, to 135. The Home Secretary was given power to fix national standards. A Central Fire Brigades Advisory Council, consisting of the main professional and governmental organisations, was formed to advise the Home Secretary on national policy, and the FBU's position as a key player in the fire service was confirmed when the union was given a place on this Council. Of course the FBU brought safety issues to the Council, but in the fire service it is impossible to separate health and safety matters from professional standards. Firefighters rely on their equipment, training, procedures and resources.

There was to be a national pension scheme, so that injustices like the one done to Alfred West would not happen again; there were to be national standards for attendance times, and for equipment. The 1947 settlement was a recognition of what firefighters had done during the war, and the contribution they had made to victory.

"Things were a lot better for firefighters than they had been before the war, and the FBU had come out of the war as a force in the land in a way that it had never been before," says its current general secretary, Matt Wrack. "The war was when the FBU established itself as a national organisation. We're a professional voice, and not just about terms and conditions."

He adds, "The union could have come out of the war destroyed." It probably would have done if it had not followed John Horner's advice and recruited the AFS on the basis of equality with regular firefighters. It's a debate that has cropped up regularly since then, but "Horner set the example, which we have followed, of resisting any attempt to create a two-tier workforce."

Horner, says Wrack, was

*a great visionary, and the most significant person in the FBU's
history. He built the union we have today, which can speak for the
overwhelming majority of the people who do the job, which can be
a voice for the whole of the fire service; and which therefore has to
be taken seriously in major policy matters.*

*During the war the union had grown to 71,500 by 1941.
Even after the end of the war and the return to a peacetime fire
service the union had 15,293 members – a 500 per cent increase
on the pre-war membership. The FBU built its strength
throughout the postwar period to such an extent that it was,
indeed, central to the functioning of the fire service.*

The 1947 settlement, with modifications, lasted for fifty-
seven years. In that time the UK's fire service acquired an
enviable reputation, both at home and abroad. The FBU made
professionalising the fire service one of its key aims, and in
1959 John Horner launched a programme called "A Service for
the 60s", which aimed to introduce fire safety inspections and
other specialist roles for firefighters. They would be sensible
alternatives to washing floors and polishing brass, which was
how some fire chiefs thought firefighters ought to be spending
their time when they were not out on calls.

In the 1964 general election, which saw the end of thirteen
years of Conservative government and the election of a Labour
government under Harold Wilson, Horner resigned after 25
years as general secretary, and, now long out of the
Communist Party, became Labour MP for Oldbury and
Halesowen. According to his friend, FBU activist Frank Bailey,

*If you were in the man's presence, you felt that you were with
someone worthwhile. He could talk intelligently, he could talk
historically . . . A parliamentary reporter told me, "The finest
speakers in the House of Commons were Winston Churchill,
Aneurin Bevan and John Horner. You could listen to those three
men and ignore all the others."*

He was a much respected Parliamentarian, but never a minister, probably because Harold Wilson was nervous about promoting an ex-Communist. Things would have been very different if he had gone into Parliament in 1945. Most of the senior ministers in 1964 – Wilson himself, James Callaghan, George Brown, Barbara Castle – were part of Labour's 1945 intake. Horner would have been one of the leaders of that intake, and almost certainly a key member of the 1964 cabinet, with a central role in shaping the Labour government's strategy. He lost his seat in Labour's 1970 defeat.

In 2004 the Fire and Rescue Services Act effectively tore up the postwar settlement. The Central Fire Brigades Advisory Council was abolished and nothing was put in its place. "The structures for the fire service which Horner helped to create were demolished in 2004," says Matt Wrack. "Since then we have been battling against the creation of the sort of fragmented service which existed prior to the war."

*

In 1977 there was a national firefighters' strike – a huge step for men and women who almost never strike. John Horner, by then a distinguished retired politician of sixty-six, walked past Soho Fire Station during the FBU strike. He wrote afterwards:

> *A lone fireman stood warming himself over his brazier, a clipboard in his hand, soliciting signatures from passers-by in support of the union. I signed the sheet and moved inside. My name meant nothing to him. I said absently, yet hoping to strike up a conversation: "I don't think I have been in this appliance room since the night the bomb fell."*
>
> *He looked at me. "What bomb?" he asked. He was very young. And I knew then that I was getting very old.*

Sources

Books

Bailey, Victor (ed). *Forged in Fire – The History of the Fire Brigades Union*. London: Lawrence and Wishart, 1992.

Demarne, Cyril. *The London Blitz – A Fireman's Tale*. London: Battle of Britain Prints International, 1991.

Farson, Negley. *Bomber's Moon*. London: Victor Gollancz, 1941.

FitzGibbon, Constantine. *London's Burning*. London: Macdonald, 1971.

Front Line 1940–41: The Official Story of the Civil Defence of Britain. London: HMSO, 1942.

Godfrey, Derek. *We Went to Blazes: an Auxiliary Fireman's Reflections*. London: T. Werner Laurie, 1941.

Goodall, Felicity. *The People's War*. [Place]: David and Charles, 2008.

Jackson, W. Eric. *London's Fire Brigade*. London: Longman Green, 1966.

Latchem, Cliff. "The Trip to the Bristol Blitz" in Thomas, A. R. (ed.). *The Peoples War*. Bath: Ina Books, 1994.

Lynn, Vera, with Robin Cross and Jenny de Gex. *We'll Meet Again: A Personal and Social History of World War Two*. London: Sidgwick & Jackson, 1989.

Mitcham, Samuel W. *Retreat to the Reich: The German Defeat in France, 1944*. [Place]: Stackpole, 2007.

Mortimer, Gavin. *The Longest Night – Voices from the London Blitz*. London: Weidenfeld and Nicolson, 2005.

Nixon, Barbara. *Raiders Overhead*. London: Scolar Press, 1980.

Parker, Peter. *The Last Veteran*. London: Fourth Estate, 2009.

Radford, Frederick H. *Fetch the Engine – the Official History of the Fire Brigades Union*. London: FBU, 1951.

Spender, Stephen. *Citizens in War – and After*. London: Harrap, 1945.

Wallington, Neil. *Firemen at War – The Work of London's Firefighters in the Second World War*. [Place]: Jeremy Mills Publishing, 2005.

Articles

"Sir Aylmer Firebrace", *Dictionary of National Biography*
The Guardian, 30 December 2004 [Harry Errington]
The Independent, 18 February 1997 [John Horner]

Websites

Canadian Firefighters in Britain World War II
<http://www.firehouse651.com/posten/index.html>
Historic Coventry
<http://www.historiccoventry.co.uk/>
Exeter Memories
<http://www.exetermemories.co.uk/em/exeterblitz.html>
Greater London Industrial Archaeology Society
<http://www.glias.org.uk/news/224news.html>
Jazz Professional
<http://www.jazzprofessional.com/Francis/As%20I%20heard%20it
%20Part%201.htm>
Norwich Evening News
http://www.eveningnews24.co.uk/content/eveningnews24/norwich
-news/story.
aspx?brand=ENOnline&category=News&tBrand=ENOnline&tCatego
ry=news&itemid=NOED11%20May%202007%2009%3A53%3A21%3A
497
Encyclopaedia of Plymouth History
<http://www.plymouthdata.info/index.htm>

Film

Filmed interviews at the London Fire Brigades Museum in Southwark.

Primary sources

Interviews with firefighters and their sons and daughters (see Acknowledgements), documentation at the London Fire Brigades Museum and London Metropolitan University, and *Firefighter*, magazine of the Fire Brigades Union, 1939–45 at London Metropolitan University.

Index